The Smoking Life

ILENE BARTH

The Genesis Press • Columbus, Mississippi

To the young: Don't even think of smoking.

To the hooked: Enjoy it while you can.

THE SMOKING LIFE
PUBLISHED BY THE GENESIS PRESS

Copyright ©1997 by Ilene Barth
The Smoking Life

All rights reserved.

This book may not be reproduced or utilized, in whole or in part, by any means. Published in the United States of America by The Genesis Press, Inc. For information: The Genesis Press, 406A 3rd Ave. N., Columbus, MS 39701-0101.

ISBN: 1-885478-22-4

Book Design by LuAnn Palazzo

Manufactured in the United States of America

First Edition

Grateful acknowledgment is made of the following previously published material:

Excerpt from *Black Elk Speaks* by John G. Neihardt, by permission of the University of Nebraska Press. Copyright ©1932,1959,1972, by John G. Neihardt. Copyright ©1961 by the John G. Neihardt Trust.

Excerpt from *For Whom the Bell Tolls* by Ernest Hemingway, reprinted with permission of Scribner, a division of Simon & Schuster. Copyright ©1946 Ernest Hemingway. Copyright renewed 1968 by Mary Hemingway.

Excerpt from *Still Life with Woodpecker* by Tom Robbins. Copyright ©1980 by Tom Robbins. Used by permission of Bantam Books, a division of Bantam Doubleday Dell Publishing Group, Inc.

Excerpt from *Hello, I Must by Going: Groucho and His Friends* by Charlotte Chandler. Copyright ©1978 Charlotte Chandler. Used by permission of Charlotte Chandler.

Excerpt from *Social Studies* by Fran Lebowitz, published by Random House. Copyright ©1980 Fran Lebowitz.

Excerpt from *Thank You for Smoking* by Christopher Buckley, published by Random House. Copyright ©1994 Christopher Buckley.

The lyrics quoted in this book are also excerpts; the words of no song are quoted entirely. Grateful acknowledgment is made for the following:

Lyrics from SMOKE RINGS by Ned Washington and H. Eugene Gifford, Copyright ©1932. Renewed by Music Sales Corporation (ASCAP) and EMI Music Publishing (ASCAP). International copyright secured. All rights reserved. Reprinted by permission.

Lyrics from YOUR SMOKE SCREEN, written by David Barbe. Copyright ©1997 Chunderman Music (BMI)/Administered by BUG Music. All rights reserved. Used by permission.

Genesis Press also requested lyric reprint permission but received no reply from the presumed owners or agents of the following songs: SMOKE! SMOKE! SMOKE THAT CIGARETTE! written by Tex Williams and Merle Travis and believed to be governed by Unichappel Music; FLOWERS ON THE WALL written by Lewis DeWitt and believed to be governed by Unichappel Music; SMOKIN IN THE BOYS ROOM written by Michael Koda and Michael Lutz and believed to be governed by Walden Music; CIGARETTES AND COFFEE written by Jerry Butler, Eddie Thomas and Jay Walker, believed to be governed by Warner Tamarlane; SMOKE DREAMS written by John Klenner, Lloyd Shaffer and Ted Steele, believed to be governed by EMI Publishing; MY LAST CIGARETTE written by Boo Hewerdine, Gary Clark, Neil McCall and Dizzy Heights, believed to be governed by Chrysalis Music; SMOKE GETS IN YOUR EYES and I WON'T DANCE written by Jerome Kern and Otto Harbach, both believed to be governed by Polygram International, Aldi Music Company and Cotton Club Publishing; CIGARETTES, WHUSKY AND WILD, WILD WOMEN written by Tim Spencer and believed to be governed by Warner Chappel; CRAZY BABY written by Rodney Crowell and believed to be governed by Warner Chappel.

The creators and/or owners of all photographs, illustrations and artwork used in this book are credited on the page on which the image appears.

All artwork noted as having come from the New York Public Library (NYPL) specifically derives from the Arents Collections, and thanks are owed to the Astor, Lenox and Tilden Foundations for their support of the library collections. The author is indebted to the curators of the Arents treasury of tobacco-related manuscripts and art for their care of these extraordinary materials.

The advertising art attributed to the Smithsonian Institution comes from its Warshaw and Ayer Collections, Business Americana, Archives Center, National Museum of American History. Most of the old movie publicity photos reproduced in this work come from the Film Stills Archive of the Museum of Modern Art (MOMA) in New York City. The author thanks the Smithsonian and MOMA curators for their guidance. The author also hails Norman Currie of the Corbis/Bettmann Photo Archive for his assistance.

Genesis Press has endeavored to identify, contact and credit the rightful owners of all literary and lyrical excerpts and art used in the book. If there have been any errors or omissions, please contact the publisher so corrections may be made in the next edition.

This book owes much to
Richard, Morgan, Kate and Daisy
who love me although I smoke.

My thanks also go to:

Laura Allen	Eric Himmel	Jonathan Ruch
Dale Berg	Lucian Hiner	Charlotte Sheedy
Rosalind Coleman	Cynthia Jenner	Victoria Smith
Ben Fleming	Jon Mindel	Charlotte Stillman
Dorothy Globus	Nicholas Morgan	Bernadette Wheeler
Richard Gollner	Sam Rehnborg	Peter Yarrow
Paul Hansen	James Rolleston	Bernie Zysberg

'It was *The Front Page* but I wouldn't have minded an ashtray.'

Would You Care to Join Me?

I started this book in love and anger. I love tobacco despite what I now know. And I've loved smokers since I was born.

My parents smoked; so did their friends. If my mom ran out we'd speed to town for two packs of Pall Malls. She'd send me into the drugstore so she wouldn't waste time parking; chocolate cigarettes were my tip.

Smoking was an alluring part of adulthood, like stockings and heels. Tobacco was romance and Johnny Mathis' tones were smoke rings for the ear. Our elders would have been relieved if they thought the Beatles only took nicotine.

I bought my first pack of Newports at age fifteen. At my ivy-draped college, there were ashtrays in the classrooms. Cigarettes cost a quarter in the dorm vending machines.

When I started working on a newspaper in 1976, the air was hazed blue-gray, and veteran reporters crushed their butts on the linoleum floor. It was *The Front Page* but I wouldn't have minded an ashtray. Now I'd take the whole mess back. Little did I imagine that carpets would come to a new city room with sealed windows, and that smokers would be pushed into the elements to enjoy the simple pleasure of our forebears.

But of course, the pleasure isn't simple; it's dangerous. The new Prohibitionists can't point to anything we don't already fear.

Fifty-five million Americans buy tobacco. Countless others are so intimidated they puff on the sly. The "nonsmokers" who recently have asked me for a cigarette include a poet, a cardiologist, a mountain climber and an elementary schoolteacher. There's something wrong with a culture so sanctimonious it turns accomplished men and women into stealth smokers.

But many in my orbit don't smoke or no longer smoke. Among their virtues is their tolerance for those who differ.

Unfortunately, there's an anti-smoking segment out there with zero tolerance. I spent over a decade either trying to give up smoking or apologizing for it. But now I, and millions like me, have been pushed too far: we can't light up in most places; strangers bark at us on public streets.

Tobacco has a history too lively to be left to the Big Tobacco companies, who 'R' not us. Forcing Corporate Tobacco to disclose what goes in its products and admit the hazards of what comes out is the most admirable achievement of the contemporary phase of the anti-tobacco movement.

The first big surprise that nicotine history yields is that anti-smoking fervor is almost as old as the export of tobacco from the New World to the Old. Nearly four hundred years ago, every bonus of tobacco had been named (and several more imagined) and almost every health hazard had also been called.

In England, tobacco was denounced by King James I in 1604. From the beginning naysayers *despised* tobacco use because it was the stuff of the American "savage." It was *other*; it was the weed of the heathens.

Tobacco sped through Europe and radiated from Lisbon to Asia and Africa. Everywhere it was introduced, it was instantly adored but often quickly outlawed—and always for the same reason. It was *other*. Sultans damned tobacco because it was unknown in the Koran. Chinese and Japanese emperors called tobacco the poison of Christian devils.

The tobacco haters may wave the flag of health, but few prohibitionists of any era take a keen interest in other people's happiness. They are too busy shouting that godliness is on their side.

While there's no longer doubt that tobacco injures a significant portion of its admirers, the evidence is weaker that secondhand smoke—ETS or environmental tobacco smoke is the highfalutin term—harms passersby. But these days, someone who, say, lights a cigar at an outdoor cafe is treated as if he just sprayed bullets on a kindergarten class.

Smokers deserve some respect.

When is Science going to return to our side? Where are the medical breakthroughs that will unchain smokers yearning to breathe free?

And when is Corporate Tobacco going to give us a break? "Good" tobacco is still a pipe dream, although formulae for "safer" cigarettes are locked in Big Tobacco vaults. Why? Marketing a cleaner cigarette would admit that what's out here now is dirty. For Big Tobacco, the concern is not smokers' well-being, but legal vulnerability.

The health Fascists clamor for cigarettes without nicotine, without real tobacco. They don't care what we inhale, as long as we don't enjoy it.

Smokers' rights? The curtailment of the liberty of a minority by a self-satisfied majority should never be shrugged off.

Pipe smokers, cigar lovers, those in thrall to sexy cigarettes—-we're part of what has always been a romantic adventure. Our heritage is noble; we follow in the moccasin steps of amazing women and men. Some who smoked before us faced troubles, too. Their noses were slit by a Mogul ruler; they were sent to Siberia by a mad czar.

Some of my best relatives are aghast at this project. I tell them: Celebrating smoking is a hard job, but someone's got to do it—-honestly. Here goes.

Ilene Barth
New York City

TABLE OF CONTENTS

Smoke Signals	7
Tobacco Makes the World Go Round	17
The Authority of the Peace Pipe	29
Nicotiana, Queen of Style	35
The First Time	45
The (Almost) Immortals	61
The Wisdom of Smokers	65
Quixotic Quitters	71
Never-Evers	83
Tobacco Fields Forever	85
The Right Snuff	101
Piping Up	111
The Once and Now Cigar	121
Chew This Over	143
Lighting up the Screen	149
Smoking Guns	163
Literary Puffs	171
Smoky Notes	185
Ad-ing It Up	197
Warning Label	221
Illusions We Lived With	229
Smokers' Luck	233
Tobacco Treasures	245
Charmed Leaf	261

'Smoking was the first politically correct act of modern times.'

God is in the details; this detail is from a 1557 Andre Thevet woodcut of Native Brazilians.

NEW YORK PUBLIC LIBRARY

Smoke Signals

FROM THE NEW WORLD TO THE OLD

All schoolchildren today know that Columbus didn't discover America. How could he? America was already there. And so was tobacco. What did you think was in those peace pipes?

On November 6, 1492, Christopher Columbus wrote in his log, "On the way inland, my two men found…men and women, carrying firebrands in their hands and herbs to smoke, which they were in the habit of doing."

This is the first written description of tobacco. The place is Cuba, although Columbus mistook it for Cathay.

Actually, Columbus had logged the existence of "dried leaves greatly esteemed by Indians" at his first landfall—wherever that was— after Indians offered the leaves along with other fruits of the earth to their surprise visitors. But Columbus hadn't a clue what to do with the esteemed leaf, until his scouts came back with their novel report.

The duo who made frequent sightings

of the smoking herb on Cuba were Rodrigo de Jerez and Luis de Torres. Jerez' previous travel experience had been in Africa, so he had no particular idea what Chinese habits might be. Torres, a baptized Jew, conveniently spoke Hebrew, Arabic and Chaldean.

Still, Columbus' description is a fair one. He even intimated that the mysterious herb might be habit forming.

Jerez picked up the habit and brought it home with him. Legend has it that when he lit up a cigar on the streets of his hometown, astounded villagers went screaming to the nearest priest. The first non-Indian smoker was soon up against the first anti-smoking group, the Spanish Inquisition. The Inquisition did take prisoners: brave Rodrigo de Jerez spent three years in jail for adopting the heathen's weed.

Bartolome de las Casas, who sailed with Columbus on his next West Indies voyage, described the "firebrand" in more detail. It was composed, he wrote in *Historia de las Indias*, of dried herbs wrapped in a leaf tube, which users "lit at one end, and at the other they chew or suck, taking in smoke" so they do not feel fatigue.

Las Casas also supplied the Taino tribal word for these organic muskets: "These tubes they call *tobacos*."

Stumbling on *tobacos* was one of the major benefits of making landfall in the Americas—no matter where you thought you were.

All hail Thomas Harriot, who knew where he had been!

Harriot was planted on Roanoke Island as part of the misbegotten Virginia colonizing launch of 1585, masterminded by Elizabeth I's favorite, Sir Walter Raleigh. After a thoroughly rotten year, with tobacco as his only solace, Harriot hitched a ride home with that peerless pirate, Francis Drake.

Soon after returning, Harriot told all in *A Brief and True Report of the Land of Virginia*, a bestseller in its day.

Harriot's star description was of "an herb which is sowed apart by itself and called *Uppowoc*," or as the Spanish heard it, *tobacos*. Harriot continued: "The leaves being dried or brought into powder, they take the fume thereof by sucking it through pipes made of clay."

The herb was a miracle, testified Harriot, which safeguarded the Indians from nasty English infections and cleared the passages and purged "gross humours." The native Virginians believed that tobacco's "sweet savour" helped quiet storms and pacify angry gods.

As a scientist, Harriot was no slouch. He went on to track Halley's Comet in 1609. But the guy wasn't a self-promoter. The comet

Sir Walter Raleigh gave tobacco its good name.

didn't get his name—although Halley hadn't even been born yet—and Harriot didn't get credit for turning Europe into a smoking zone. Nope. Walter Raleigh stole his fire.

Walt had enjoyed the ear (and maybe more) of Queen Elizabeth ever since he'd tossed his cape over "a plashy place" so her Stiff-Neckedness wouldn't muss her shoes. After Thomas Harriot returned to merrie England, the Bess-Walt affair heated up.

Everywhere the proud queen went, Sir Walter was sure to follow, dragging on his silver pipe. And somewhere under his cloak (now cleaned and dried) was his gold tobacco box.

Smoking became the rage of the Elizabethan A-list. Paul Hentzner, a Dutchman enjoying London in 1598, remarked, "The English are constantly smoking." From their pipes, he wrote a friend, "they draw the smoke into their mouths which they puff out again through their nostrils, like funnels, along with it plenty of phlegm from the head."

The dandies in the better seats of the open-air Globe Theater puffed from ornate pipes as a playwright named Will Shakespeare strutted his stuff. But the masses in the pit had to make do with walnut shells attached to straw stems.

Elizabethan ladies also piped, although maybe not in public. But when you remember that Juliet was played by a cross-dresser, you realize it's pretty hard to say what a lady was, never mind what she did.

Did the queen smoke? Dramatist Thomas Middleton took up this high-collar question in *The Roaring Girle*, presented less than a decade after Elizabeth's death. In it, Moll Cutpurse avers that Her Royal Highness tried it: "After two or three whiffs she was seized with a nausea. Upon observing this some of the Earl of Leicester's faction whispered that Sir Walter had certainly poisoned her. But her majesty, in a short while recovering, made the Countess of Nottingham and all her maids smoke a whole pipe among them."

Plausible. But would you believe any-

thing someone named Moll Cutpurse says?

The real Moll Cutpurse, née Mary King, was a notorious Elizabethan courtesan who dressed in men's clothes and was said to have made her living as a fortuneteller, pickpocket and forger. She died in 1659, at the age of seventy-five, but according to a contemporary "would probably have lived longer had she not smoked tobacco."

Cutpurse's rep aside, another rumor with legs is that Queen Elizabeth was a gambler. One day she bet Sir Walter that he could not weigh his beloved smoke. He showed her he could. He tapped a pinch of tobacco onto scales, then put the stuff in his pipe and smoked it. Finally, he weighed the ashes. The difference, he proclaimed, was the weight of smoke. Her majesty paid up with the comment that she'd heard of many who turned gold into smoke, "but Raleigh was the first who had turned smoke into gold."

The first Iron Lady drew the line at stuffing tobacco up your nose for religious purposes; she forbade snuff-taking at church. Elizabeth I passed from the scene in the year 1603. She was the original patron of smokers' rights, and after she died tolerance for tobacco diminished in olde London towne.

Her successor, James I, formerly James VI of Scotland, lost no time in issuing his 1604 *Counterblaste to Tobacco*, which called smoking a "custome loathesome to the eye, hatefull to the nose, harmefull to the braine, daungerous to the lunges and in the black stinking fume thereof, nearest resembling the horrible Stigian smoke of the pit that is bottomlesse."

That was the good news. Jimmy Stuart's main objection to the herb that had taken his new kingdom by storm was that it reeked of witchcraft—the sorcery of "savages" clouding over faire England and Scotland.

The battle was joined.

The king locked Sir Walter Raleigh in the Tower for several years, let him out to look for gold, and claimed his head when he failed to find it. Raleigh approached the scaffold after his breakfast pipe and asked his executioner for permission to feel the axe edge. "This is a sharp medicine but it is physician for all diseases," Sir Walter observed. "When I stretch forth my hands, dispatch me."

John Aubrey wrote of Raleigh's last pipe, "Some female persons were scandalized, but I think it was well done and properly done to settle his spirits."

The trumped-up charges against the late queen's main man didn't mention smoking—they called it treason.

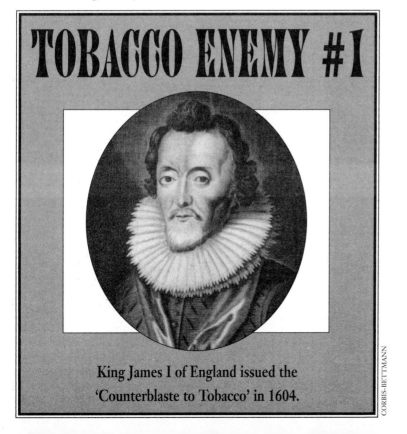

King James I of England issued the 'Counterblaste to Tobacco' in 1604.

As an absolute monarch, James I could do whatever he wanted. While Raleigh still lived, the king arranged a debate at Oxford in 1605 to settle—once and for all—the question of whether tobacco was bad for you. He forced his own doctor, Sir William Parry, who was a steady smoker (and, incidentally, lived to eighty) to argue against the witchy weed. Eyewitnesses reported that the reluctant physician admirably presented his master's case, but that tobacco's brave defender, Dr. John Cheynell, was no slouch, either.

The only absolute verdict reached at Oxford was that it was better not to smoke when King James was around.

Edmund Gardiner, a "gent and practitioner in Physicke" followed up in a pamphlet entitled *The Triall of Tobacco*, inventing this fine epitaph for a smoker:

> Here lieth he had lived longer if
> He had not choaked himself with
> a Tabacco whiff.

The opposition rallied behind the wisdom of one Roger Marbecke, another medical man, who before the counterblaste had described the aroma of a good smoke: "how divine it is, how much more like the rose."

Another smoker's friend was Dr. Raphael Thorius, who responded to the fuss with a carol of tobacco praise, *Hymnus Tabaci*, which labeled the herb "a banquet for the Braine."

In Latin and English, the Brits were puffing day and night, and not only the men. Encounters with Jacobean ladies pulling from delicate clay pipes were recorded for posterity.

Perhaps reflecting that hell has no wrath like smokers scorned, the king reacted as politicians usually do when confronted with rampant vice. He taxed it.

Thanks to Pocahontas, Captain John Smith lived to smoke another day.

Under James I, the duty on a pound of tobacco went up four thousand per cent. When his American subjects heard about his taxes, they laughed up their deerskin sleeves. Fledgling tobacco barons did what they had to do: they sailed their cargo around customs officers who couldn't be persuaded to tip the scales in their favor. Business boomed although there would be no significant tax-relief until Queen Anne, of club-foot-furniture fame, so decreed in 1713.

Smoking was the first politically correct act of modern times. Among its champions were the most famous bi-cultural duo in American history, Pocahontas and Capt. John Smith, leader of the Jamestown settlement, named after

Tobacco Enemy Number One.

Let's look again at that day when Chief Powhatan, wielding a stone war club, pinned John Smith to the ground. The captain was done for until the chief's 11 year-old daughter, Pocahontas, dived into the historic scene, throwing her arms around the captive's head.

Did you ever ask your second-grade teacher how Captain Smith got in that fix? Of course, not. Well, let me tell you: Smith was plotting the murder of the good chief, a premeditated homicide so he could steal the Indians' corn. And, to get in the mood, Smith had detained—okay, let's call it kidnapped—Powhatan's little girl. But Pocahontas bore no grudges, and she was the apple of her father's eye.

And so Captain Smith lived to smoke again—-the pipe that may have been his was unearthed from the original Jamestown site in 1996.

Pocahontas went on to marry shipwreck survivor John Rolfe, developer of the hybrid tobacco that turned North America into a treasure chest.

Did Pocahontas die of first or second-hand smoke? She did not. In London, she was carried away by that scourge of Old World civilization, smallpox. Rolfe, returning to Virginia without his princess, was slain by the natives. But their son, and Smith (retired and scribbling in England), lived on to tell the tobacco-crossed tale.

Although the Jamestown fort burned down, thousands of acres in the Virginia colony (including the territory that would one day separate as North Carolina) were given over to the great cash crop. In 1616, the first ship whose entire cargo consisted of tobacco arrived in England.

His Majesty collected his sin tax (or part of it), and sinners on both sides of the Atlantic continued to inhale.

In 1621, Robert Burton in *Anatomy of Melancholy* crooned, "Tobacco, divine, rare super-excellent tobacco, which goes far beyond all the panaceas, potable gold and philosophers stones." But Burton also saw the down side, lamenting "tis a plague, a mischief, a violent purger of goods, land, health, hellish devilish and damned tobacco, the ruin and overthrow of body and soul."

Four years later, Londoner William Cramden noted that "tobacco tavernes" had become more popular than alehouses.

Meanwhile, New World Puritans—-the crowd but lately fetched up on Plymouth Rock in the name of freedom—outlawed public smoking in the Bay colony. But no one was exiled to the great north woods with a scarlet "S" emblazoned on the bosom.

Connecticut Yankees were even more weirded out over smoking than their Massachusetts brethren. A 1650 blue law decreed that only a traveler venturing at least ten miles might smoke on the journey—-but only once. Three centuries later, this rule became the airlines' model.

Not all colonials north of the Mason/Dixon line were wary of tobacco.

Banned in Boston

Pennsylvania was founded by a smoker, and William Penn partly paid for land with three hundred pipes. The Quakers, male and female, puffed peacefully in the city of brotherly love.

And, truth to tell, it wasn't long before the pilgrims' descendants grabbed whatever shred of the tobacco trade they could.

Back in England, Puritan Oliver Cromwell groaned. But although he had been able to overthrow the monarchy, he could not topple tobacco. When he died in 1658, his Roundhead soldiers celebrated by smoking during the funeral.

Five centuries have passed since Native Americans first shared their herb of choice with Europeans, but human temperament remains the same. For every questing Columbus and valiant Pocahontas, there's someone who hates their guts.

THE SMOKING LIFE 15

'The Old Man of Palenque'

400 A.D.
This Mayan stone carving from Chipas, Mexico, is the oldest artistic rendition of tobacco smoking although a 4000-year-old pipe has been unearthed from a Wyoming mound.

Tobacco Makes the World Go Round

A SMOKING TIME LINE

CATHERINE DE MEDICI -- FIRST LADY OF SNUFF

AGE OF INNOCENCE · 1492 · 1499 · 1528 · 1534 · 1560

1492 — Christopher Columbus discovers tobacco.

1499 — Amerigo Vespucci witnesses a new sport, tobacco spitting, on what's now the Venezuelan island, Margarita.

1528 — Hernando Cortes, having conquered Mexico for Spain, gives the Portuguese king Aztec tobacco seeds to plant in his Lisbon garden.

1534 — Explorer Jacques Cartier describes Iroquois petum use—in pipes with lit charcoal on top—but tobacco doesn't captivate the French until...

Age of Innocence — Native Americans cultivate a most fragrant weed, honored and enjoyed on both their northern and southern continents.

1560 — **CATHERINE DE MEDICI,** *mother of the French king, suffers one of her royal headaches. She turns to the leaf, a brain remedy, presented to her by Jean Nicot who learned of tobacco's curative powers in Portugal. The queen daintily stuffs pulverized tobacco up her royal nose to clear her head. Soon toute Paris clamors for the queen's asprin, christened nicotine in honor of the gift giver.*

THE SMOKING LIFE

SAYONARA TOBAKO

King James of England issues his "Counterblaste to Tobacco."

Put opium in your pipes and smoke it—and leave tobacco to the Christian round-eyes. So warns an edict of the waning Ming dynasty of China.

1595   1604   1605   1607   1612   1615

THE "BARBARIAN" GIFT *comes to Japan. But the emperor of the sun orders shoguns to stomp on the European habit.*

Arab merchants buy tobacco from the Portuguese for their water pipes.

The Portuguese ply West Africa with Brazilian tobacco; the Africans like it so much that a traveling man named Harris observes, "they go hungry rather than do without" their tobacco.

When in Rome, light up: Cardinal Crescenzio carries his nicotine habit from London to the Vatican.

Jahangir, Mogul Emperor of Hindustan, orders smokers' lips split. Jahangir's problem? Tobacco isn't mentioned in the Koran.

TURKISH DELIGHT

SHAH SEFI *outlaws tobacco in Persia to the surprise of the accidental French tourist Jean-Baptise Tavernier who writes, "men and women are so addicted that to take tobacco from them is to take their lives." Sefi does just that. The punishment for two men arrested for smoking: Pour molten lead down their throats until they are dead. At last Sefi dies. Long live his son, Abbas II. The first important act of Abbas II, a committed chain smoker, is the repeal of his father's tobacco ban.*

1617 **1620** **1632** **1634** **1640**

James I flows with the smoke and sells pipe-making licenses. The Society of Tobacco-Pipemakers' motto: Let brotherly love continue.

THE TURKS *so adore tobacco they grow it instead of poppies, and imbibe its fumes—"four pillows on the couch of pleasure"—from hookahs. But ruler Murad IV takes a heady dislike to their pleasure. He skulks around Constantinople in mufti looking for tobacco connections. When he scores, the seller is decapitated.*

Moscow's burning! And Czar Mikhail Fedorovich orders smokers' noses sliced. That doesn't work. He orders them executed. That also doesn't work.

ONE MEAN AYATOLLAH

THE SMOKING LIFE **19**

Pope Urban VIII enjoins women, laymen and especially priests from smoking while celebrating mass.

Dutch traders convince Capetown locals to add tobacco to their dakka (hemp) pipes and find that African women prefer the new brew.

BORIS, IT'S COLD OUT HERE

1642   **1649**   **1650**   **1652**   **1679**

CZAR ALEXIS *sends smokers to Siberia. That doesn't work either.*

Ireland goes green as smutchin (snuff-taking) sets the nation sneezing.

German apothecary Godfrey Haukwitz manufactures sulfur-dipped sticks, which he sells with phosphorus paper—the first matches.

Peter the Great lives up to his name, declaring himself a smoking devotee. All Russia has been dragging ever since.

TOBACCO THINK TANK

Royal Tobacco Factory of Seville is Spain's biggest employer. Their prize products are cigars, but pinches of scrap tobacco get rolled in paper to produce the *papelito* or *cigarito*.

1689   1725   1740   1750   1789

FREDERICK THE GREAT of *Prussia revives the Tabaks Collegium to advise him on matters of state.*

Pope Benedict XIII rules that snuff may be taken in St. Peter's to stop the early exits of those "who can't abstain from tobacco use."

From the *sans culottes*, the folks who brought us Bastille Day, comes the popularization of the *cigarette*. The cheap *cigarito*, with its name gone French, becomes the people's answer to the snuffed noses of the aristocrats.

THE SMOKING LIFE   21

1790 — *Mano a Mano:* The hands that make Cuban cigars add bands, a labeling first to fight-off market frauds.

1831 — Turks add stiff mouthpieces to small rolled smokes—the filter tip is born.

1832 — Posselt and Reimann isolate nicotine from tobacco in Germany.

1836 — A Frenchman on holiday in Corsica breaks his meerschaum and begs a local craftsman for a replacement. The best the Corsican carver can do is fashion a pipe of briar. Soon the woodturners of Saint Claude, France, are manufacturing the pipe that won the west.

1839 — Stephen "Slade," an 18 year-old slave, drops charcoal on a North Carolina curing fire and creates the great granddaddy of modern tobacco, mellow "Bright."

1843 — *Voila!* The French are the first to promote cigarette brand names. But Gitane and Gauloise will not debut until 1910.

Abraham Lincoln, who teethed on corn pipes, taxes cigars to finance the Civil War. So both North and South fight on tobacco money.

EMMA CALVE AS CARMEN, 1898.

THAT'S ENTERTAINMENT!

James Bonsack perfects cigarette-rolling machine, spewing out 200 smokes per minute.

1856   **1864**   **1865**   **1875**   **1884**

GEORGES BIZET debuts "Carmen," his opera of a gypsy cigarette girl in Seville.

"A heart in love is quickly burned,
It knows no law except its own desire.
If I should love you and you spurn me,
I'm warning you, you play with fire!"

A Bond Street tobacconist, specializing in Havana cigars, fashions Turkish cigarettes to satisfy cravings of returning Crimean vets. Tiny Philip Morris, Ltd. will morph into the world's biggest smoke purveyor.

Washington Duke of Durham, N.C., stripped of all else by Gen. Sherman's boys, sets out with his nine-year-old son Buck to peddle a hunk of prewar chew. So begins the American Tobacco Co.

THE SMOKING LIFE **23**

TOBACCO ENEMY

NORMAN ALLEY, 1913

LUCY GASTON 1860 - 1924

Seven billion cigars and five billion cigarettes sold in U.S. And Richard Joshua (R.J.) Reynolds introduces a pipe tobacco named after Queen Victoria's oldest son, spawning the century's dumbest joke: Do you have Prince Albert in the can? Yes. Why, don't you think you ought to let him out?

1899

ILLINOIS SCHOOLMARM *Lucy Gaston becomes the Carry Nation of nicotine when she launches the Chicago Anti-Cigarette League in 1899. Eleven years later she goes national. Her belle lettre, "Coffin Nails," becomes her platform in her 1920 run for president of the United States. Although her White House quest fails to make most history books, Gaston persuades victor Warren Harding not to smoke in public. Lucy Gaston, the only politician never to enter a smoke-filled room, dies at the age of 64—of throat cancer.*

1900

Illinois Supreme Court finds state ban on cigarette use unconstitutional— but 19 other states experiment with tobacco prohibition before World War I.

1907

1911

Sherman Anti-Trust Act busts American Tobacco, which controls four out of five smokes in the U.S. and England.

Auntie Em can light up legally, as Kansas becomes the last U.S. state to repeal its cigarette ban.

Presto—the Zippo lighter is born.

STOGIE STORK

THE BABY BOOM BEGINS and *New York tobacconist Nat Sherman peddles cigars with color-coded ribbons— pink or blue—for the first time.*

1926   1927   1928   1932   1937   1945   1945

Racecar driver Spud Hughes invents the mentholated

U.S. cigarette sales bang the gong at 100 billion.

The first king-size cigarette, 85 mm Regent, appears.

French women get rations for cigarettes and the right to vote.

THE SMOKING LIFE **25**

1950
Researchers statistically link smoking to cancer: the Ernest Wynder-Evarts Graham study is published in the "Journal of American Medicine;" the Richard Doll-Bradford Hill paper appears in the "British Medical Journal."

1962
British cigarette makers agree not to advertise on radio or the telly before 9 P.M.

1964
The report of U.S. Surgeon General Luther Terry concludes "cigarette smoking is a health hazard."

1966
Warning labels appear on U.S. cigarette packs.

1970
Cigarette ads outlawed on American radio and TV

1981
American cigarette market peaks at 640 billion smokes.

1987
Prince Albert escapes! After eighty years in the can, R. J. Reynolds marries Nabisco and sells off the little prince. John Middleton, Inc. bags Albert and seals him in a red box.

The French greet smoke-free zones in restaurants with a Gallic shrug.

Turkey issues tough anti-tobacco rules. A Turk explains, "We want smoking stopped throughout the world." Murad IV wasn't that ambitious.

Cigarette maker, Liggett, admits nicotine is addictive and smoking can cause cancer. U.S. gets court okay to regulate nicotine as a drug. Big Tobacco offers to hang the Marlboro Man and other icons, while paying $368 billion in reparations—to government.

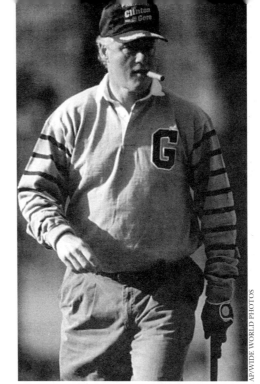

LEADER OF THE NOT SO FREE WORLD

1992 1996 1996 1996   1997

FIFTY-FIVE *million Americans smoke as President Bill Clinton and opponent Bob Dole debate additional restrictions. Clinton, who threw smokers out of the White House, nonetheless pulls the Smoker vote.*

California and Texas researchers report that BPDE, a chemical in tobacco smoke, causes cell mutation identical to that in malignant lung tumors.

As one billion smokers light up worldwide, China is leader of the pack.

THE SMOKING LIFE **27**

"Is not the sky a father and the earth a mother, and are not all living things with feet or wings or roots their children?"

Black Elk, 1931

The Authority of the Peace Pipe

When Caribbean tribesmen gave Columbus and his men tobacco, it was a peace offering. The Genovese captain and his seamen dressed oddly, but they traveled in style and their hosts were welcoming. Perhaps the Indians also wished them godspeed on their journey elsewhere, *anywhere but here.*

It wasn't until explorers of the 1500s were blown to more northerly latitudes that Europeans encountered the peace pipe or *calumet*. There were also war pipes but it was the ceremonial pipe of peace that hailed pale visitors.

Jacques Marquette's striking description of the peace pipe was published in France in 1673. Marquette was a Jesuit missionary who, with Louis Jolliet, had explored the upper Mississippi and happened on Illinois and Wisconsin. This is what impressed him:

"I must speak here of the Calumet, the most mysterious thing in the world. The sceptres of our kings are not so much respected. The Savages have so much respect for this pipe that one may call it *The God of Peace and War*, or *The Arbiter of Life and Death*.

"One, with this Calumet, may venture amongst his enemies, and in the hottest Engagement they lay down their Arms before this Sacred Pipe.

"The Illinois presented us with one of these which was very useful in our voyage. Their Calumet of Peace is different from the Calumet of War: they make use of the former to seal their Alliances and Treaties, to travel with safety and receive Strangers; the other is to proclaim War."

The origin of the peace pipe was

interpreted by another white man, Henry Wadsworth Longfellow, in his 1840 epic, *Song of Hiawatha*.

> On the mountains of the Prairie,
> On the Red Pipe-stone quarry,
> Gitche Manite, the mighty,
> He the Master of Life, descending
> On the red crags of the quarry
> Stood erect, and called the nations,
> Called the tribes of men together...
>
> From the red stone of the quarry
> With his hand he broke a fragment,
> Molded it into a pipe head,
> Shaped and fashioned it with figures;
> From the margin of the river
> Took a long reed for a pipe stem
> With its dark green leaves upon it;
> Filled the pipe with bark of willow,
> With the bark of the red willow;
> Breathed upon the neighboring forest,
> Made its great boughs chafe together,
> Till in flame they burst and kindled;
> And erect upon the mountains,
> Gitche Manito, the Mighty
> Smoked the calumet, the Peace Pipe,
> As a signal to the nations.

Longfellow drew on Native American lore, but no one has spoken more eloquently of the peace pipe than Black Elk, an Oglala Sioux holy man on the Pine Ridge Reservation in South Dakota.

Black Elk was not explaining some exotic custom in 1931 as he handled the peace pipe. He was describing its ancient but still living meaning. His wisdom, excerpted here, was recorded by Nebraska poet John G. Neihardt.

BLACK ELK SPEAKS: THE OFFERING OF THE PIPE

"I will first make an offering and send a voice to the Spirit of the World that it help me to be true. See, I fill this sacred pipe with the bark of the red willow; but before we smoke it you must see how it is made and what it means.

"These four ribbons hanging here on the stem are the four quarters of the universe. The black one for the west where the thunder beings live to send us rain; the white one for the north, whence comes the great white cleansing wind, the red one for the east, whence springs the light and where the morning star lives to give us wisdom; the yellow one for the south, whence comes the summer and the power to grow.

"But these four spirits are only one Spirit after all, and this eagle feather here is for that One, which is like a father and also is for the thoughts of men that should rise as high as eagles do.

"Is not the sky a father and the earth a mother, and are not all living things with feet or wings or roots their children?

"And this hide upon the mouthpiece, which should be a bison hide, is for the earth, from whence we came and from whose breast we suck as babies all our lives, along with all the animals and birds and trees and grasses.

"And because it means all this, and more than any man can understand, this pipe is holy.

"There is a story about the way the pipe first came to us. A very long time ago, they say, two scouts were out looking for bison. When they came to the top of the hill and looked north, they saw something coming a long way off. And when it came closer, they cried out, 'It is a woman!' and it was.

"When she came out of the cloud, they saw that she wore a fine white buckskin dress, that her hair was very long and she was very beautiful. She spoke: 'You shall go home and tell your people that I am coming and that a big teepee shall be built for me in the center of the nation.'

"The people did at once as they were told, and around the big teepee they waited for

The Great Spirit

the sacred woman. And after a while she came, very beautiful and singing, and as she went into the teepee this is what she sang:

> With a visible breath I am walking.
> A voice I am sending as I walk.
> In a sacred manner I am walking.
> With visible tracks I am walking.
> In a sacred manner I walk.

"As she sang there came from her mouth a white cloud that was good to smell. Then she gave something to the chief, and it was a pipe—-with a bison calf carved on one side to mean the earth that bears and feeds us, and with twelve eagle feathers hanging from the stem to mean the sky and the twelve moons, and these were tied with a grass that never breaks.

" 'Behold!' she said. 'With this you shall multiply and be a good nation. Nothing but good shall come from it.'

"Now I light the pipe, and after I have offered it to the powers that are one Power, and sent forth a voice to them, we shall smoke together.

"Great Spirit you have been always and before you nobody has been.

"Hear me four quarters of the world, a relative I am.

"Great Spirit, Great Spirit, my Grandfather, all over the earth the faces of living things are alike. With tenderness have these things come up out of the ground. Look upon these faces of children without number and their children in their arms, that we may face the winds and walk the good road to the day of quiet.

"This is my prayer; hear me. The voice I have sent is weak, yet with earnestness I have sent it. Hear me.

"It is finished. *Hetchetu aloh!*

"Now my friend let us smoke together so that there may be only good between us."

'Real men smoked cigars, dandies preferred cigarettes and bookish types liked pipes. And what about the ladies?'

Nicotiana, Queen of Style

The fragrance, the ember's glow, the sensual blue-filtered world—-there is a paradise of reasons to love tobacco, but how we love is a reflection of our time and place.

Nicotiana is a mistress of disguises. Humans have invented a myriad of ways to imbibe tobacco. Every single one of them Native Americans thought of first.

Tobacco was more than a recreational drug for Native Americans. It was a holy plant. Tobacco might be the wings of an individual vision quest or the heartbeat of tribal ceremony.

American societies also employed tobacco to combat cowardice, illness, injury, fatigue. A 16th-century Spanish physician named Nicholas Monardes catalogued tobacco's curative powers in *Of the Tabaco and its Greate Vertues*. No medical man's work was so widely read or copied until Dr. Spock came along.

Taken as Europe was with the medicinal properties of tobacco, the Old World also worshipped the herb for the psychological enlightenment smoking brought, and was ever open to novel means of instant nicotine absorption. The sublime tobacco experience was that which combined feeling better with getting better.

In New York's American Museum of Natural History there's a small exhibit, easy to miss. It features an Amazon tribe's apparatus for pulverizing tobacco and the shaman's instrument for administering the result. The shaman's tool has two prongs, one for each nostril. It is an elegant device for taking powdered tobacco, a.k.a. snuff.

Shamans self-dosed, the better to

communicate with higher Spirits. They also made their stash available to others—for purely medicinal purposes. Medicine women and men prescribed snuff to clear the head of whatever: migraines, gloomy thoughts, tumors.

Inhalable snuff tobacco chased European smokers off the fashion runway.

After Jean Nicot told the French Queen Mother that snuff would take care of her awful headaches, French aristocrats snuffed their noses nonstop. The French oozed *je ne sais quoi*. Maybe the sneeze soiree doesn't sound so enchanting, but if the French were doing it, it was stylish. And such chic was not lost on the courtiers and courtesans in far-flung capitals. Besides, those little jeweled snuff boxes were so cunning. Snuff became the tobacco choice of aristos across Europe.

Enter the plague. The year is 1665 and it's nasty out there. But the rumor surfaces that there's a way to protect yourself without fleeing to a desert island (hellishly hard to find when you need one). The way is to dam your nasal passages with snuff. A London bourgeois noted this in his diary. The diarist, natch, was Samuel Pepys.

So it happened that in much of Europe snuff floated from the royals to the aspiring classes. The 1700s were Snuff Century. Ah-choo.

Taking snuff was thought dainty; smoking was not. A German lady rebelled against being consigned to powdering her nose. In 1715 she published *A Sound and Pleasant Proof that a Respectable Woman without Damage to her Good*

Name may and should treat herself to a Pipe of Tobacco. This, incidentally, is the abbreviated title of her pamphlet.

In mid-century, John Wesley, founder of the Methodist church, put a backhanded seal of approval on the Big Sneeze when he forbade followers to smoke.

Still the Dutch, most of the English working class and various other stubborn personalities stuck to their pipes. In 1773, Samuel Johnson complained to his biographer, James Boswell, "Smoking has gone out. To be sure, it is a shocking thing blowing smoak out our mouths into other people's eyes, mouths and noses, and having the same thing done to us. Yet I cannot account why a thing that requires so little exertion and preserves the mind from total vacuity should have gone out."

Boswell was only moved to celebrate the stylishness of the nose candy:

Oh Snuff, our fashionable end and aim!
Strasburgh, Rappee, Dutch, Scotch,
   whate'er thy name
Powder celestial! Quintessence divine!
New joys entrance my soul while
   thou art mine.

Dr. Johnson became a snuffer, too.

Such tribes as survived the European onslaught kept up their traditional tobacco practices. But the colonists and, after 1776, the new Americans, followed European fashion—-with one exception. The great frontier democratizer was chew. Why bother with smoking implements, or take the time to rasp tobacco to a fare-thee-well, when you're beating back the wild?

To finance the new republic Alexander Hamilton proposed a snuff tax; the political point was this would not apply to the working person's tobacco.

Snuff's snooty reign crept into the 19th-Century. In 1834, a New York *Literary World* writer proclaimed, "Old Brazilian Indians were the fathers of snuff and its best fabricators. Their taste was as pure as that of the fashionable world of the West."

Actually, powdered tobacco had become a sign of the style-challenged. The bloom was gone from the nose in Europe; New York just hadn't heard yet.

In France, where the craze had begun, snuff use became an invitation to the guillotine during the French revolution. In the name of *fraternité* and *égalité*, the populace had taken to cigarettes, the cheap smokes modeled on Spain's *papelito*. Not that the Spanish had originated the cigarette; credit might more properly go to the Aztecs, who had

enjoyed tobacco shavings stuffed in hollow reeds.

France's hold on the title, Tobacco Fashion Capital, slipped away when Napoleon Bonaparte persisted in being a snuffhead as the world around him smoked.

The newer story of the 19th Century was the cigar, the rich boy toy—outside the Hispanic world where it was commonplace. In the early 1800s, English gentlemen could obtain cigars only from the West Indies, so of course they wanted them. And the gold-braid officers of several countries took to cigars, so much handier than pipes or snuff on the battlefield, even if you never left headquarters.

In 1814, English troops fought the French in Spain. The truce revealed agreement on this point: Cuban cigars were the best smoke in the world. The after-dinner cigar became de riguer in European salons.

Cigars did not so quickly capture North America. By and large, the sneezing stopped, but chewable snuff took over. The frontier moved into the U.S. Senate, courtrooms, clubs, city streets. When Charles Dickens visited Washington, D.C. he marveled at the projectile culture. Spittoons were the new American sculpture; unfortunately several public venues were unadorned.

During the Crimean War (1853-1856) West met East with one result more outstanding than Florence Nightingale. English officers embraced Russian and Turkish cigarettes. They returned home so heady with the experience that

FUNCKING THE CORSICAN.

they demanded that their local cigar rollers replicate the pleasure.

Napoleon III, Emperor of France, showed himself a cigarette devotee, but this did not seal the rise of the cigarette. Across the ocean, Ulysses S. Grant was making his cigar famous—-that time lag thing again.

Up Yours!

Tobacco was a staple in the medicine cabinet of Native Americans. It was used topically to heal wounds, erase poison from arrow punctures and calm rashes. Remedies for internal infection depended on the ingestion of tobacco leaf or on the intake of smoke, and not only through the mouth and nostrils.

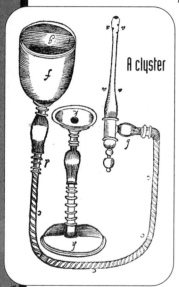

A clyster

The borrowed tobacco cure which ignited the ingenuity of European doctors was the smoke purge. Guaranteed to fix indigestion and wipe out other bad humors of the digestive tract, kidney stones included, the challenge was how best to stream smoke into the appropriate orifice.

Indians had worked the smoke enema by the expedient of having the bare-bottomed patient squat over an outdoor tobacco fire. If the patient was too ill to achieve the proper balance, mates held him on a straw stretcher over the flames. But the European medical establishment found drawbacks to these methods: Both fire and patient were subject to wind and rain, too much of the cure might be blown away or a good gust could set the patient on fire. All in all, some bluebloods found the treatment a trifle inelegant.

In 1611, the Danish physician Thomas Bartholin prescribed "smoke from two pipes blown into the intestines" to alleviate upsets. And he thoughtfully urged a stylish instrument to accomplish this: the clyster. The clyster was improved upon by several of the best scientific minds of the time, until it achieved a stunning Rube Goldberg complexity, which seemed to satisfy the carriage trade.

Tobacco historian George Apperson noted, "The cigar dealt the snuff box its death blow and the cigarette was the chief mourner at its funeral."

The second half of the 19th century was an era of style splits. Real men smoked cigars, dandies preferred cigarettes and bookish types liked pipes. And what about the ladies? Oh dear.

You see, Queen Victoria disapproved. Completely. She loathed cigars, pipes, cigarettes. She thought smoking unbecoming to a gentleman or a lady. Some silly rules were made in the Admiralty but, by and large, gentlemen ignored Queen Victoria.

Actually, no one ignored Queen Victoria to her face. Even the Prince of Wales, a famous chain smoker of cigarettes and cigars, dared not light up in his mother's presence. Victorian prudery was powerful. The lights went out for women throughout Britain, North America and elsewhere. A German etiquette book advised females that their proper smoking role was embroidering tobacco pouches.

There were rebels. The French novelist George Sand scandalized a few visitors with her beloved cigars. And an anonymous Brit huffed in 1869: "When one hears of sly cigarettes between feminine lips at croquet parties, there is no more to be said."

Victoria ruled for almost sixty-four dark years. Her son did not assume the throne until 1901 when he was over sixty. The coronation of Edward VII was a brilliant affair. His first words after the dinner following his ascension were, "Gentlemen, you may smoke."

Young women welcomed the new regime, too. A few years after the century's turn, this verse was a Punch and Judy bit:

> A pretty, piquant, pouting pet,
> Who likes to muse and take her ease
> She loves to smoke a cigarette.
>
> A winsome, cool, clever coquette
> Who flouts all Grundian decrees—-
> A pretty, piquant, pouting pet
> Who loves to smoke a cigarette.

Although there had been grumbling against cigars in the 19th century, it was the cigarette that inflamed neo-Puritans. Cigarettes were modern, breezy, inexpensive—-the preference of artists, women and young men.

Animus against cigarettes had been expressed as early as 1889 in this *New York Times* editorial: "The decadence of Spain began when the Spanish adopted cigarettes and if this pernicious practice obtains among adult Americans the ruin of the republic is close at hand."

Closer than Timesmen feared. From 1902 on, sales of pipe tobacco, cigars and chewing tobacco declined while cigarette sales rose.

More than one pipe-smoking parson on both sides of the Atlantic announced that the demon-cigarette was the destroyer of boys. Health concerns

A Real Prince. When he became King Edward VII, he declared, "Gentlemen, you may smoke."

THE SMOKING LIFE

were also voiced in a few popular magazines. Influential robber-baron Henry Ford self-published *The Case Against the White Slaver* in 1914. The alleged enslaver was the cigarette, not the assembly line. Ford attributed the cigarette's peril to the burning paper. He ominously warned, "If you will study the history of almost any criminal, you will find he is an inveterate smoker. Boys, through cigarettes, train with bad company. They go to pool rooms and saloons."

Ford's book congratulated his pal, the cigar-smoking Thomas Alva Edison, for his decision not to hire cigarette smokers.

Youth was not deterred. A year after Ford's book, this student doggerel appeared in a Penn State publication:

Tobacco is a dirty weed. I like it.
It satisfies no normal need. I like it.
It makes you thin, it makes you lean,
It takes the hair right off your bean,
It's the worse darn stuff I've ever seen.
I like it.

But American politicians, who would soon Prohibit alcohol, were not immune to anti-cigarette pressure. Several states banned sales to minors; Indiana forbade possession by anyone of any age. Crusader Lucy Gaston looked like a winner until World War I sent remnants of Victorian propriety up in smoke. Fighting boys demanded and received their tobacco due. After the war, women slipped out of long dresses and into cigarettes.

From the art deco films of the '30s to the motorcycle glam of '50s-'60s flicks, cigarettes dominated Euro-American tobacco chic. Among its classy admirers were Franklin Delano Roosevelt, Coco Chanel, Jacqueline Kennedy and Princess Margaret.

Although the cigarette took a status dive in the west as hard bodies and aerobic lungs became fashion icons, it has remained the populist choice. Meanwhile in Eastern Europe and Asia, American filter-tip cigarettes have become a style statement. What could be more symbolically prosperous than the svelte Virginia-tobacco smoke?

However, Anglo-American trendies prefer the cigar. Only a few years ago, a cigar was thought a no-class, dirty smoke. Now it has risen from its ashes as a power smoke—-the nouveau choice.

The cigar is just a phallus by another name, insists an unimpressed friend. She disagrees with Sigmund Freud's analysis that "sometimes a cigar is just a smoke."

Women need no longer envy men their cigars. Cigar dinners, cigar clubs, expensive cigar paraphernalia are chic for both genders.

We will greet the new millennium with multi-mannered smoke.

"Davidoff," 1997, by Arianna Caroli

'You can hide de fier, but wat you g'wine do wid de smoke?'

Uncle Remus
Plantation Proverbs by Joel Chandler Harris

The First Time

"Who was ever delighted with tobacco the first time he took it?" mused Sir Kent Digby in 1630 or thereabouts. "And who could willingly be without it, after he was for a while habituated to the use of it?"

Tobacco is less harsh today but for many the first go is still jolting. For others it's love at first whiff. Seduced or repelled, the initial tobacco encounter is memorable.

Person: Kitty Barnes
Place: Chicago
Year: 1960

"I was thirteen and had a boyfriend who was three years older. He was tall and ele-

gant and well-mannered, and he smoked Kents. I think I thought if I smoked I'd be tall and elegant, too. So I bought a pack of Kents. It was summertime and we went to Halloran Park where we sat on the grass and watched the kids play baseball. I opened the pack and my boyfriend said, 'What do you think you are doing?' I said, 'Smoking,' as if I did it every day. I lit a cigarette and literally saw stars. The rush made me fall backwards. But my boyfriend didn't notice, falling backwards wasn't obvious since we were sitting on a grassy knoll. It was a near-death experience but the truth is I loved it, the smoking, and have for the rest of my life."

Person: *Sylvia Carter*
Place: *Farmhouse outside Granger, Missouri*
Year: *1951*

"My dad was known as Happy Carter when he was a kid, and everything he did looked desirable and delicious to me. One thing he did was chew 'tabakker.' When I was about five, he'd leave his plug on the top of the dresser. Now I was always climbing up to places I wasn't supposed to be. One day I climbed up to get a taste of this yummy stuff. I took a great, lusty taste—-and knew right away it wasn't to my liking.

"I spit it out. But even under duress, I didn't spit it on the rug. I'd been trained you took great, good care of the carpet. So I either spit it into my hand or raced to one of the slop buckets in the kitchen. Then I thought it over. Trying tobacco seemed like a good joke. So I gave some to my friend Becky and told her it tasted just like candy. Becky didn't like it, either."

Carter Reprise: New York Daily News, 1969

"I decided to smoke for real. The reporters I admired smoked. It would keep me awake. I thought of smoking as a New York kind of thing. On the day I decided this I went for broke. I went down to the newsstand and bought a pack of Gauloise. The pack was French blue, the blue of Colette's blue paper. I was an addict right away—-two to three packs a day.

"It was pretty tough to smoke that many Gauloise each day. It's like drinking brandy all through the day. So after a while I decided

Gauloise should be kept for feast days; that my everyday smoke would be Camels. Camels are sweet. I once saw a clipping that explained that one of the reasons Camels are so mild and so smooth compared to other filterless cigarettes is they have caramel and chocolate floor sweepings as fillers."

Carter Last smoke: Kansas City, 1985

"I had given up cigarettes again and had just spent several hours driving to the stockyards for a story. I had lunch with the president of the stockyards. We ate steak and black coffee—-that's what you eat there. Then we went back to his office where he opened up a drawer of his desk and took out a glass fruitjar of Arkansas moonshine. I said, 'I'd love to drink with you, I don't think I can without a cigarette.' He was trying to cut down so he said, "You're welcome to one of mine but it's a sorry excuse for a cigarette.' It was a Now. I took it and smoked it and drank his white lightning. He was right; it was a sorry excuse for a cigarette."

HADDOCK, PHILA., CIRCA 1875. NEW YORK PUBLIC LIBRARY

Person: **Murray Kempton**
Place: **Baltimore**
Year: **1934**

"I was eighteen, in college, and had been on cigarettes about a year, a pack a day, and it was getting to me. So I decided to switch to a pipe. I bought a corncob for a quarter. I remember that it made me sick. But I tried again and it was fine and we've lived happily together ever since. When I look at the price of cigarettes, I'm glad I made the change.

"I have a different pipe now [1996], but I use the same tobacco, Edgeworth, which is a plain, slow-burning Burley. It's getting harder and harder to find. I hate the aromatic tobaccos; they cloud my head. I used to smoke a pound of tobacco each day but now there are so many places you can't smoke I'm down to half a pound. I don't think there's any question but that my lungs look as if I've worked in a textile mill all my life, and every October there's bronchitis. I can do some things without my pipe but I don't think I could work without smoking.

"You know I have a friend, he was the Czech foreign minister for about five and a half minutes, and he was here [Manhattan] visiting. We took a lunchtime walk and he said, 'Look at all the prostitutes.' I had to tell him they weren't prostitutes; they were perfectly respectable working girls standing in doorways, sitting on ledges, having a cigarette. The girls, poor things, have nowhere else to smoke."

Person: **Hal Price**
Place: **Houston**
Year: **1985**

"Smoke made me cough when I was a kid, but when I reached thirty my life was mostly work—-I own a hair salon—-and stress. And everyone was smoking with cocktails. So I took up cigarettes, at first in social situations.

"I think smoking is an outgrowth of the drug culture. The people who used other [substances] are dead; I'm still here. I'm health conscious. I work out every day. I bike and ski.

"I did quit once, with acupuncture, but I went up to 210 pounds in nine months. So I thought this is crazy and I started smoking again.

"I have to run because I'm illegally

parked."

—Price darted into the drugstore, emerged with Marlboro Lites and jumped into his car.

Person: **Matt Brubaker**
Place: **Panama City, Florida**
Year: **1992**

"I'm nineteen now [1996] and I started during a lunch break from ninth grade. We were across the street from the school, outside a deli, and I took a Marlboro off a guy. It hit me like a heart attack. I took a drag—and whew!

"But I liked people seeing me with a cigarette. And I came to like the headrush, the dizzy feeling.

"Then I got caught smoking behind my church. This girl saw me. She told the leader of my youth group who told my parents who grounded me for a week. So I had to smoke in my room—-the old fan behind the window trick."

Person: **Sir James Matthew Barrie**
Place: **Kirriemuir, Scotland**
Year: **1873**

"I was a schoolboy living with my brother who was a man... [One evening] we sat together in the study. I cannot say at what time I had an inkling that there was something wrong. I heard people going up and down stairs, but I was at that time not actually suspicious.

"[Yet] I felt my brother had something on his mind. As a rule, when we were left together, he yawned or drummed his fingers on the arm of his chair to show me that he did not feel uncomfortable. I made a pretence of being at ease by playing with the dog. In this crafty way we helped each other. On that occasion he did not adopt any of the usual methods and though I went up to my bedroom and listened through the wall I heard nothing.

"At last someone told me not to go upstairs and I returned to the study. He was still in his armchair and I again took the couch. I could see by the way he looked at me over his pipe, he was wondering if I knew anything. But the affair upstairs was too delicate to talk of...

"He lit his pipe and pretended to read. Every five minutes his pipe went out, and sometimes the book lay neglected on his knee as he stared at the fire. Then he would go out for five minutes and come back again...We sat defiantly.

"At last he started from his chair as someone knocked at the door. I heard several people talking, then loud above their voices a younger one.

"The question was, what was the proper thing for me to do? I told myself my brother might come back at any moment. What should I say to him? I heard him coming down. When his hand touched the door I snatched at my book and read as hard as I could.

"He was swaggering a little as he entered but the swagger went out of him as his eye fell on me...At length he sat down again and took up his book. The silence was something terrible; nothing was to be heard but an occasional cinder falling from the grate. This lasted I should say for twenty minutes, and then he closed his book and flung it on the table. I saw that the game was up.

"He said with affected jocularity, 'Well young man, do you know that you are an uncle?'

"After a while I said, in a weak voice, 'Boy or girl?'

" 'Girl,' he answered.

"I thought hard again and all at once remembered something. 'Both doing well?' I whispered.

" 'Yes,' he said sternly.

"I felt that something great was expected of me but I could not jump up and wring his hand. I was an uncle. I stretched out my arm toward the cigar box and firmly lit my first cigar."

Person: Lou Schwartz
Place: Syracuse, New York
Year: 1950

"I didn't smoke in high school but I when I got to university I hustled myself a job distributing free Chesterfields. The kids who were giving away the samples also had to check with campus outlets to see what was the best seller. Chesterfield came in third. When we reported this to the Chesterfield people, they dressed us down as if it were our fault. So the next time we just made up the numbers and put Chesterfield first. The company was so

LIGGETT & MYERS, 1941

pleased with this information, it used it in its advertising: Chesterfield is Number One at Syracuse University.

"About that time, I started smoking them myself."

Person: *Monica Ernst*
Place: *Caracas, Venezuela*
Year *1993*

"I was married then but I had a boyfriend who liked cigars. I'd never smoked—-tobacco, that is—-and really wasn't interested in it. But I thought I'd surprise my boyfriend with *Romeo y Julieta*; that's what Winston Churchill smoked, and besides the name was so romantic. It was going to be a gift.

"But before I had a chance to give the cigars to my boyfriend, my husband went rummaging in my big bag for something and came up with the cigars. I said the first thing that popped into my head: I told him I'd taken up smoking.

"He glared at me. So I opened up a cigar—I'd watched my boyfriend get one ready—-and lit it. But I inhaled and couldn't

"THE TRIUMPH OF TOBACCO."

JOHN WATLACE, 1837 SPECIAL COLLECTIONS LIBRARY, DUKE UNIVERSITY

stop coughing.

"I'm no longer married and I don't have the same boyfriend. But once in a while I have a cigar at a club. If you can handle it, and I can now, guys think it's sexy."

Person: **Ben Fleming**
Place: **Bredyard, Sweden**
Year: **1992**

"I was twenty-four and had left Australia to travel, and a young lady named Anna was showing me her town. We'd spent the day looking at red and white painted 'storybook' houses, old churches, the local peat bog and walking through an ancient cypress forest. Then Anna, her boyfriend, Pers, and I had something to eat and some beer. Afterwards, Pers asked me if I'd like some *snus*.

" 'What's *snus?*' I asked. Anna and Pers laughed and whispered to each other in Swedish. 'I'll show you. You'll like it,' Pers said to me.

"*Snus* are like little one-third size teabags but filled with dried tobacco leaves. The *snus*, or dip, is inserted onto the top gum of the mouth and held in place by the top lip. The idea is to spit out your saliva into a little cup while enjoying post-dinner conversation.

"I tried it. The tobacco went through my gums into the bloodstream and straight to my brain. It was an awesome nicotine rush.

"The downside is if you don't spit out your saliva, pure droplets of nicotine shoot to your stomach. That's what happened to me. Five minutes after my hit, Pers found me over the toilet, throwing up."

Person: **Wendy Brooks**
Place: **Hackensack, New Jersey**
Year: **1954**

"I was outside Hodde's Ice Cream Parlor with four high school sorority sisters in a '53 Bel Air. Everyone was smoking but me. 'Why don't you have a drag?' the girls said. I was thinking, I'll have to try or I won't be able to stay in this group. So I took a cigarette.

"I coughed and kept on coughing. I was so embarrassed; it proved I was unworthy. I took a few more drags and let it burn down between my fingers.

"We went into Hodde's and I smoked two more. I just surrendered to it.

"I'd grown up in a cloud of smoke and always liked the smell of tobacco. It reminded me of World War II, of sitting on the sofa with my mother and grandmother who were darning sox and smoking

"But my father was an athlete and he didn't like smoking so my mother quit. After that day in Hodde's it took me a long time before I'd openly buy a pack.

"For years smoking was okay. But then I rode a bus into San Francisco and I saw this big sign that said: Secondhand Smoke Kills. It had never occurred to me that my smoking had any relationship to anyone else. Maybe there was some truth there. That sign has haunted me ever since. The knowledge didn't alter my behavior but it altered my psyche. I started feeling guilty."

Person: Samuel Taylor Coleridge
Place: Cambridge, England
Year: 1790

"On the assurance that the tobacco was mild, I took half a pipe, filling the lower half of the bowl with salt. I was soon, however, compelled to resign it, in consequence of a giddiness and distressed feeling in my eyes...Soon after, deeming myself recovered, I sallied forth to my engagement...I had scarcely entered the minister's living room, ere I sank back in the sofa in a sort of swoon...My face was like a wall that is white-washing, with cold drops of perspiration running down it from my forehead...I at length awoke from insensibility, my eyes dazzled by the candles which had been lighted in the interim."

—The poet took to opium, instead.

Person: Jim Mitchell
Place: Richmond, Virginia
Year: 1994

"I was at a testimonial dinner at the Commonwealth Club, honoring the president of our company for fifty years of service. And he was a cigar smoker. I was over forty but tobacco had never interested me when I was young.

"After the dinner and before the speeches, brandy and cigars were served. I'd been to similar dinners before and I'd always said 'no' to the cigar. But this was special. I was

honored to be at this dinner. The man being honored was a nice gentleman, and I was captured by the moment. I took a cigar and I liked having it in my mouth.

"Someone took the time to educate me. A friend said: just let the cigar smoke roll in your mouth.

"It's a nicotine delivery system with the advantage that you don't send the smoke through your lungs. I buy Honduran cigars; it's a mild, smooth smoke and lasts about an hour. Smoking is just a great outdoor activity, especially when I'm playing golf."

 Person: Lea Smith
Place: Friedburg-on-Hess, Germany
Year: 1965

"I was an exchange student for the summer. And it was lonely for me because the family with whom I was living was cold; I think they just wanted me there as their English teacher.

"Everyone in Europe smoked and in the afternoons I just had to escape. So I took to going out, buying German cigarettes and settling down with a book. The only English-language novels available were by P.G. Wodehouse and Ian Fleming. So I spent the summer smoking and reading James Bond.

"When I got back home I started going out with my girlfriends and smoking. It was so boring in Des Moines that there was nothing else to do but drive around and smoke.

"When my father saw me, even though he was a smoker, he went bananas. I was grounded for most of my senior year."

Smith Reprise: Washington, D.C., 1992

"I hadn't had a cigarette for ten years. I'd gained thirty-five pounds and my husband would say to me: You're just not the same not smoking. But quitting had been hell and for ten years I was adamant about not starting again.

"It was the night of Clinton's first inaugural and I was whirling through parties with my sister who was a smoker. Everyone smoked. All the campaign workers were kids who lived on cigarettes, coffee and McDonalds. Finally, at an MTV party, I couldn't stand it any more. I took a cigarette. I loved it. I was so happy. I was an instant addict again. The next morning I stole some of sister's cigarettes before I started home."

Person: **Doris Anderson**
Place: **Mylan, Illinois**
Year: **1943**

"I got a job in a factory. It was during the war, you know. The only trouble was I was only fifteen and you were supposed to be sixteen to work in that place. So I bought myself a pack of Kools so I'd look older.

"But one day I left them behind in the house and my mother found them. She went around to the neighbors to see who they belonged to. When I came home, I fessed up. And my mom had a fit. She made me eat the entire pack.

"I got very sick. I also got so mad I vowed to smoke.

"But I never did inhale until the first time I got drunk. I was with my boyfriend and I was eighteen when that happened."

Person: **Nina Pinsky**
Place: **San Francisco**
Year: **1957**

"My friend, Tina Bernal, and I used to steal Kents from my parents and smoke

them in the basement, next to the old coal bin. Or we'd go to Tina's and smoke in the garage. This was when we were nine years old.

"One day Tina's house burned down. And the fire started in the garage! Tina and I were terrified and swore to each other that neither of us would tell.

"A few days later, this little seven-year-old hoodlum who lived in the neighborhood admitted to starting the fire. He'd been playing with matches in the Bernal garage and had run out when things started burning."

"So the scare didn't deter us from smoking again. But then my father decided he wanted to turn the old coal room into a wine cellar and my mother came down to take a look. She found our tray of butts there and was furious.

"I didn't smoke again until college. I don't know about Tina. She became a nun."

Person: Richard Gollner
Place: Skegness, England
Year: 1959

"I was thirteen and at the seaside for the first time in my life, and there was a slot machine there. You put in one penny and you either lost, or won a cigarette. I got a Park Drive. It was the shortest and cheapest cigarette you could get, but I found it sweet.

"I found I could blow smoke rings and girls loved it. I was told if I practiced hard I'd be

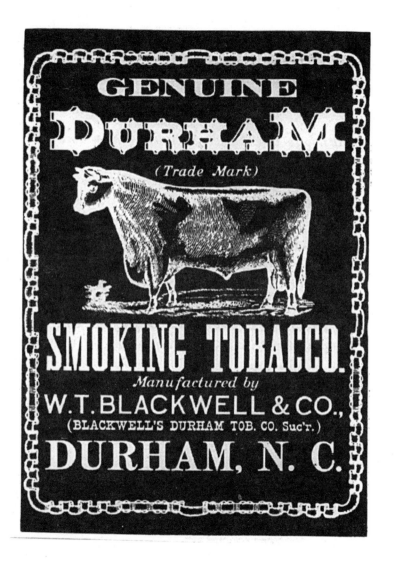

able to blow smoke hearts, but I've never accomplished that."

Person: *Dick Fountain*
Place: *Montgomery County, Pennsylvania*
Year: *1947*

"I was a counselor at a Boy Scout camp, Camp Belmont, the summer I was fifteen. It was around a campfire. I didn't care for it my first try, but all my friends were smoking so I tried it again. Then I smoked anything I could lay my hands on. After a while Bull Durham was my brand; I rolled my own. A few years later I joined the Navy, spent three years, mostly in Alaska. Cigarettes were only sixteen cents a pack so I smoked constantly."

Fountain reprise:

"I quit for five years. But one day I took a sniff and I was hooked again. That was twenty years ago. I'll go for anything as long as its natural burley. I don't like the over-the-counter brands with shellac as a preservative."

People: *Judy Garcia and Emily Diaz*
Place: *New York City*
Year: *1992*

Emily: "When I was twelve, I was out at a Dominican restaurant in Queens with my sister and some friends. We were waiting for the food. My sister, she was twenty then, goes: 'Do you wanna cigarette?' Yeah. I wanted to try it. Like everyone else was smoking."

Judy: "It was a Newport, right? My first was a Newport, and I was twelve, too."

Emily: "Yeah, it was a Newport."

Judy: "In Junior High I used to go to this Saturday sports program. And the kids would bring stuff to drink— you know, beer— into the locker room. So I was drinking and decided to try a cigarette. It tasted nasty but I got used to it."

Emily: "After dinner, I had another one and fell dizzy on the floor. Everyone was laughing but my sister helped me up and got me water. Then I had another cigarette and did okay."

Judy: "I do a pack a day now. Or as many as I can get. Still Newport."

Emily: "Sometimes I have to spend my lunch money. And once in a while I think I'll quit. Before I get old. I know about the cancer. I don't want to lose my hair."

Judy: "But there are times you really need a cigarette. Like after sex."

Smokers Are Sexy

A recent book, "Smoking: the Artificial Passion," by California academic David Krogh labels smoking "expensive, dirty and deadly." But after reviewing psychological studies of smokers, Krogh acknowledged:

Smokers have what can only be called a higher sex drive.

Smokers are more honest than non-smokers in the view of themselves they present to others.

'I said it was a matter of honor, remember? They called me a chicken. You know, chicken? I had to go 'cause if I didn't I'd never be able to face those kids again.'

James Dean
Rebel Without a Cause, 1955

The (Almost) Immortals

One* out of four smokers will die of a tobacco-related disease. Three out of four won't.

EIGHT SMOKERS AND HOW THEY DIED

Sir Walter Raleigh.... Lost his head.

Amelia Earhart.......... Disappeared into thin air.

Catherine the Great..... Fell under a horse.

Abraham Lincoln.......... Shot in the third act.

James Dean.................. Crashed.

John Fitzgerald Kennedy................ Went to Texas.

*Winston Churchill............. Stroked out at age 90.

*Pablo Picasso.................. Heart broke when he was 91.

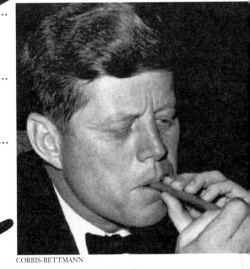

THE SMOKING LIFE 61

WORLD'S OLDEST SMOKERS

The annals of tobacco history exult in legendary smokers whose lives reached the century mark. American comedian George Burns, a.k.a. Oh God, reached nico-stardom decades before he chuckled out in 1996, several weeks after his hundredth birthday. Hollywood immortalized Burns' El Producto Queens print in the cement outside Mann's Chinese Theatre. Old George had a good run although he

George Burns had a good run.

never came within a cigar's length of winning the longevity sweepstakes.

The year is 1769, but who's counting? The Swiss are—-when Abraham Fauret bids his pipe and everything else good-bye at the age of 104. Fauret, who was born in 1665, spent much of his life with his head in an oven. He was a baker. To the last, it has come to us, this Swiss citizen walked firmly with a pipe in his mouth.

Serpent's Tale

Tobacco got the Islamic heave-ho because it wasn't in the Koran until Arabic storytellers smoked up this legend:

The Mohammed was taking a stroll when he saw a serpent, stiff with cold, on the ground. He took it up and warmed it in his robes. When the snake had recovered, it said, "Divine prophet, I am now going to bite thee."
"Why, pray tell?"
"Because thy race persecutes mine."
"How can you be so ungrateful?"
"There is no such thing as gratitude on this earth. By Allah, I shall bite thee."
"If thou has sworn by Allah I'll not cause you to break thy vow." The prophet held out his hand to the snake who bit it. Mohammed sucked up the venom, spat it on the ground and from this sprang tobacco, combining the poison of the snake and the compassion of Allah.

Fauret's Gold Medal was snatched away in 1856 by one Jane Garbutt (the Olympic Smokers' Marathon is a coed event) of North Riding, Yorkshire, who entered for England. She was 110 years of age when she puffed her last. Obituaries, written at the time, agree that she had smoked a pipe for one hundred years.

But she may have been outdone by Peter Garden, who is said to have "lived to the uncommon age of 131 years-old with his faculties entire to the last." The trouble is, we don't know who said this, and on what authority. We have Garden's likeness, but the artist and his subject might have lived in Brigadoon, the Lerner and Loewe musical community that comes to life only one day each century. That kind of life keeps a laddie young, no matter how old he is—-or how much he smokes.

The undisputed Smoking Immortal of modern times is Jeanne Calment of Arles, France, who was born on Feb. 21, 1875 and died on August 4, 1997, at the brilliant and impeccably-certified age of 122. She had become the oldest person in the world at age 116. Only after obtaining landmark status did this fabulous femme, who knew Van Gogh (another smoker), give up the cigarettes which had kept her young.

Mme. Calment passed the torch to Christian Mortensen of San Rafael, California, who used it to light his cigar. Mortensen, who was born in Denmark on Aug. 16, 1882, became the oldest smoker on earth just days short of his 115th birthday. He was edged out of the oldest-person title by Canadian Marie-Louise Meilleur who, having been born on Aug. 29, 1880, is nearly two years his senior. Mrs. Meilleur has about three hundred protective descendants, who have not told the world whether their foremother ever smoked.

Did the world's longest-lived smoker reside in Brigadoon?

'We must not forget that a pack of cigarettes, the ceremony of taking one out, igniting the lighter, and the strange cloud which surrounds us, have seduced and conquered the world.'

Jean Cocteau

The Wisdom of Smokers
ON SMOKING AND OTHER SUBJECTS

Louis Armstrong
"Man if you gotta ask, you'll never know."

W.H. Auden
"Of course 'behaviorism' works. So does torture."

James Baldwin
"God gave Noah the rainbow sign, No more water, the fire next time!"

Honoré de Balzac
"Tobacco is a cure for the sickness of civilization."

Ludwig von Beethoven
"I want to seize fate by the throat."

Milton Berle
"I've been smoking since the age of twelve, and that's okay because you don't inhale cigar smoke."

Otto von Bismarck
"When a man begins a discussion which may easily lead to a heated argument, it is always better to smoke as you talk."

Thomas Carlyle
"Tobacco smoke is the one element in which, by our European manners, men can sit silent together without embarrassment."

Coco Chanel
"As long as you know most men are like children, you know everything."

Lord Clarendon
"Diplomacy is entirely a question of the weed. I can always settle beforehand a quarrel if I know [what] the plenipotentiary smokes. Tobacco—the key to democracy, is my theme."

Santa Claus
"Happy Christmas to all, and to all a good night."

Charles Darwin
"Nothing can be more improving to a young naturalist than a journey in distant countries."

Adelle Davis
"Let's find out how to stay fit."

Sammy Davis, Jr.
"Yes, I can."

Walt Disney
"I'm an optimist...I can still be amazed at the wonders of the world."

George Gershwin
"True music must repeat the thought and inspirations of the people and the time. My people are Americans. My time is today."

Albert Einstein
"I shall never believe that God plays dice with the world."

T.S. Eliot
"Where is the wisdom we have lost in knowledge?"

Sigmund Freud
"I owe to the cigar a great intensification in my ability to work."

Thomas Hobbes
"The condition of man…is a condition of war of everyone against everyone."

Oliver Wendell Holmes
"I must not smoke so persistently. I must turn over a new leaf—a new tobacco leaf—and have a cigar only after each cigar."

Jack Kerouac
"The reason there are so many things is because the mind breaks it up."

Tamara de Lempicka
"I'm in the process of being rediscovered."

THE SMOKING LIFE **67**

Brave New Huxley

"For forty years of my life tobacco had been a deadly poison to me, I hated tobacco. I could almost have lent my support to any institution that had as its object the putting of smokers to death.

"A few years ago I was in Brittany with some friends. They were happy [although] outside it was wet and dismal. I thought I would try a cigar. I smoked that cigar—it was delicious. From that moment I was a changed man.

"I now feel that smoking in moderation is a comfortable and laudable practice and is productive of good."

Aldous Huxley

Madonna
"Express yourself."

Jerry Mahoney
"I'm a dummy."

Golda Meir
"I must govern the clock, not be governed by it."

Demi Moore
"I've tried the Montecristo No. 2, the torpedo. It's a little big for me but I like it." (*Cigar Aficionado*)

Edward R. Murrow
"We must not confuse dissent with disloyalty."

Isaac Newton
"If I have seen further it is by standing on the shoulders of giants."

J. Robert Oppenheimer
"Physicists have known sin and this is a knowledge they cannot lose."

Popeye
"I am what I am."

William H. Rehnquist
"In the light of temptation that naturally besets any human being who becomes a judge of the Supreme Court, the remarkable fact is not that its members may have on infrequent occasions succumbed to these temptations, but that they have by and large had the good judgment and common sense to rise above them."

Norman Rockwell
"I paint life as I would like it to be."

Babe Ruth
"I like to live as big as I can."

Arnold Schwarzenegger
"Hasta la vista, baby."

Robert Louis Stevenson
"No woman should marry a teetotaler or a man who does not smoke."

Mark Twain
"If I cannot smoke in heaven, than I shall not go."

John Wayne
"Talk low, talk slow, and don't say too much."

Simone Weil
"Purity is the ability to contemplate defilement."

Oscar Wilde
"A cigarette is the perfect type of a perfect pleasure. It is exquisite and it leaves one unsatisfied. What more can one want?"

'For thy sake, Tobacco, I would do anything but die.'

Charles Lamb, 1805

Quixotic Quitters
GIVING UP IS HARD TO DO

The quote on the facing page from *Farewell to Tobacco* has become the most famous statement about smoking ever made. Serious contemporary analysts use the 1805 declaration Charles Lamb sent to Wordsworth to preface their attacks on Big Tobacco; stop-smoking groups employ it to stun recruits. Time and again, I have subscribed to its sentiment.

Not until I was deep in research for this book did I discover that the essayist had been unable to live up to his avowal. Ten years later, Lamb swore to his friend, Thomas Manning, "This very night I am going to leave off tobacco!"

But Lamb, although he may have stopped smoking for brief intervals, was not able to leave off tobacco. As late as 1830, and perhaps longer, he was wedded to his pipe.

It doesn't cheer me that legions—I among them—have tried to stop smoking and failed. Nicotine is habit-forming, and we who use it regularly are addicts. Another essayist, Sir Francis Bacon, recognized this four centuries ago in his *History of Life and Death*. Bacon wrote, "Tobacco in this age grown so common, and yielding such a secret delight and content, that being once taken, it can hardly be forsaken."

Manufacturers of tobacco products have long known what they offer us is addictive; satisfying our craving is their business. I fault them not for this, but for their coy evasions of the truths that make them rich.

Fifteen million smokers attempt to quit annually; one million succeed. That's what one set of numbers says, but I've seen no statistics I consider indisputable. By another count,

there are forty million American ex-smokers walking around today.

The conventional wisdom is that most quitters try stopping several times before they succeed. What's certain is that smokers who don't develop the habit until they're adult are much more likely to wean themselves from it than those of us who became hooked in youth. You better believe that Big Tobacco knows that too.

People, far too many people, say that giving up smoking is a simple matter of bending one's will to it. I even heard this from a tobacco grower who no longer smokes. On the other hand, former heroin addicts have told me that tough as it is to kiss off heroin, it's easier than redemption from nicotine.

Some people claim that when they stopped smoking, they felt better immediately—could within weeks scale K2 or some such nonsense. But I've never felt bad smoking. My worst days have come from trying to give it up.

No doubt, my lungs aren't pretty. Here's another ugly truth: a person who quits after having smoked as long as I have doesn't substantially reduce her cancer risk for twenty years. But the possibility of cardiovascular disease immediately plummets. That should impress me—-and in a way, it does—-but then I remember that I have low cholesterol and low blood pressure, and I recall not how good a cigarette tastes but how much it hurts not to have one.

When I took up smoking I didn't know it was bad for me. Both my parents smoked and my sister and brother soon would, too. When I first became aware of the downside, I paid scant attention.

Thank God that my own children aren't so dumb. Even the littlest, who is only four, hates smoke.

My father smoked for thirty years, rarely inhaling and quitting when he was over fifty. He lived to be eighty-six, dying of heartbreak weeks after my mother, sixty-four, went first. The cause of her death was never pinned down, but it was likely smoke-related.

The man I married also liked cigarettes, although he barely went through two packs per week. He stopped for good when our first child was born. I was puffing only lightly then, as has been my discipline during pregnancy and nursing. For me the difference between quitting and cutting down is wider than the Grand Canyon. (When I smoked there, I packed my butts out.)

Phony Express

What do U.S. and French postal authorities have in common, besides a propensity to lose the mail? Both censor smoking on stamps. American authorities struck the cigarette from the lips of blues musician Robert Johnson. French officials issued a stamp bearing a well-known portrait of André Malraux. Missing from the famous likeness was the philosopher's dangling cigarette.

Also in the now-you-see-it, now-you-don't category is "I Love Lucy," whose initial sponsor was Philip Morris. The lead-in to the original episodes was a cartoon Lucy and Desi smoking. The hallmark was left on the cutting room floor when those segments were "restored."

My addiction shames me and I don't smoke in closed spaces with my children nearby. But I throw open windows and have been known to stand on porches in my nightgown in zero-degree weather. Working, I almost always have a smoke in hand. Writing, decision-making, even sweet nothings glide gracefully for me only when a cigarette is close by. In a "quit" phase, I'm surly, sleepless and don't get anything done.

I never met a cigarette I didn't like. France for me is the aroma of Gauloise. When I travel, I always snag the local tobacco in the manner others dig in to exotic dinners. (I do that, too.)

These days, no place is as uncomfortable for an American smoker as back home. As much as I sometimes detest my own dependency, I hate the smugness of those who would strike me and my kind from this earth.

The first time I tried to quit, everyone in my pantheon of relatives and friends was still happily hooked. I enrolled in SmokeEnders, which had me wrapping my Winstons in a rubber-band maze. I had to want to really get at one of my smokes. Then I had to write down when I smoked it, and why. Lighting up was forbidden while driving or going to the movies.

Since I then spent a huge chunk of my life on the Long Island Expressway, I soon went from two packs per day to half a pack. Some days the new regimen made me think that what I should quit was not smoking, but commuting (which I eventually did), but mostly I was proud of my progress.

As for movies, I went as often as I could, substituting a wad of chewing gum for the desire to smoke when the actors on the big screen lit up. One night, the woman in front of me turned around and hissed, "Would you stop chewing so loud?"

The program's endgame was a countdown to zero cigarettes. Key was the "buddy" each of us was assigned, a fellow traveler in the quest for kicking the habit. My fellow traveler was also a nonstop traveler: Jake was a cabbie.

Buddies were supposed to telephone each other whenever the urge surged. Jake pulled over to a curb to ring me from a pay phone twelve or fifteen times daily. He jerked me from my computer and out of editorial meetings. I ran out of the shower, awoke from deep sleeps and interrupted lovemaking to talk Jake down.

And what happened when I needed him? Nothing! Jake, zinging in his yellow taxi throughout New York City, was impossible to reach.

I went up to four and five cigarettes daily and then...

I eventually tried to stop smoking again. And again. It wasn't happening. Hypnosis tensed me up. The only suggestion I welcomed was a cigarette break. Nicorette gum tasted so nasty, I spit it out. Nicotine-laced "candy" drops were just as bad.

I was becoming a dinosaur in my world as smoking restrictions piled up and my Merit Ultra Lights became inadmissible in professional circles.

In California, my brother quit cold turkey and took up jogging and cycling. I embraced tennis and skiing—without converting. Since then, he's suffered one sports injury after another. I've been fine.

A relative, who has never smoked, was

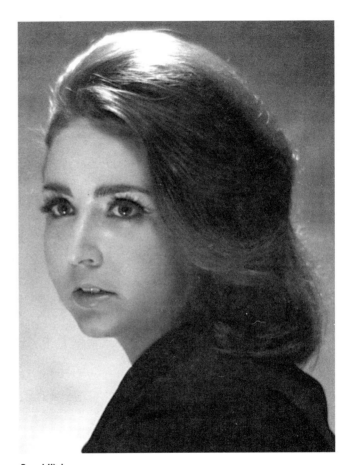

Ronni Kleinman
"There is no friend like a sister." —Christina Rossetti

stricken with cancer in her mid-forties. Life is unfair.

Seven years ago, my sister was diagnosed with an end-stage brain tumor. The cancer had started in her lungs. Her doctor didn't tell her to stop smoking; he didn't want to ruin her final months.

But I enrolled in an addiction clinic run by a psychiatrist who also treated heroin and cocaine users. After the stop date, he ran urine tests on his patients to see if they were cheating.

There was a strange solace in admitting I was a hard-core addict—right down there with the folks who had track marks running up their arms and the crackheads who'd rather get high than feed their babies. It was an acknowledgment that smoking is not a simple self-indulgence.

The weekend I was to stop, I headed with my family for the snow-covered Vermont mountains. The reasoning was the more active I could stay, the less I'd miss my fix. But when surrender-day came, I stayed in bed, unable to face the world with no possibility of ever having a cigarette again.

I tried to sleep, I clawed my pillow, drank endless glasses of water and made an ungodly but solitary fool of myself. But I didn't smoke.

The day after that was no easier. Within a few hours, my family would have happily traded in the harridan in their midst for the smoker they knew—not that they'd admit it. It never got easier for me. Or them.

A month passed. I went to my newspaper job every day. My work was terrible, but at first the people around me were understanding. I was able to smell smoke on people's breath, not that I found it unpleasant. But I was surprised to find out that my dentist smoked, as did my regular doctor, along with several co-workers who allegedly had stopped. I didn't call anyone's game; I was too shaky about mine.

Most of the time I wasn't at the paper I was with my sister who lived alone. She chain-smoked as she talked of deep matters, and I was content to share her fumes.

The cancer ward impressed me when I visited my sister after her first round of chemo. Ronni—still connected to tubes—shakily got up. I helped her into her robe and she pointed to her night-table drawer. In it were two packs of cigarettes and a lighter. I put them in her pocket.

She leaned on me, her tubes on wheels following her, as we made our way down the hall. I asked the nurse, "Where may we smoke?"

We took the elevator to the top floor, then I helped her and her apparatus up three steps, where a door opened onto an unadorned roof. She giggled at the awkwardness of it all, the first time I'd heard Ronni laugh since her diagnosis.

She pulled out her cigarettes, offering one to me. I don't know if she'd forgotten that I'd stopped or just hadn't noticed. Short-term memory isn't too reliable when a tumor is eating your brain.

I took it. I lit her cigarette, then mine. Declining would have been a cruel reproach. It would have said: You're dying and it's your own fault.

We smoked companionably on the roof of "Cancer Central."

I knew what the clinic shrink would say: You didn't have to take it. You took that cigarette because you want to smoke. If you valued your life, you'd cut free of your sister.

That's what he said. So I stopped going to the clinic.

My sister lasted six months.

I'm still smoking, but I feel less beleaguered than I did a year ago. The ranks of cigarette recidivists are growing, and cigars are chic. Restaurants that permit lighting up are thronged. It's sad, it's ironic, it's thrilling.

I haven't given up on giving up but I take solace in knowing that stronger wills than mine have wrapped themselves around quitting and failed.

Maybe smokers are an endangered species clinging together in foolish hope of survival as the ship sinks. The ship is listing. but smoking steadies us for a while.

The Final Quit awaits whether we smoke or not.

LOST BOY GROWS UP

J. M. Barrie, the creator of Peter Pan,

loved it all: pipes, cigars and cigarettes. But as a bachelor "drifting towards a tragic middle age," he decided to descend from Neverland. "To lay aside my pipe was to find myself soon afterward wandering restlessly around my table. No blind beggar was ever more abjectly led by his dog, or more loathe to cut the string," he remembered. To ease his emancipation from the "mean slavery," he wrote the affectionate tribute to tobacco, *My Lady Nicotine*, excerpted here in "The First Time."

TO SMOKE IS TO BE

In his thirties, Jean-Paul Sartre forsook his pipe. He wrote, "I used to smoke at performances, mornings at work, evenings after dinner...Whatever unexpected event struck me, it seemed to me that it was fundamentally impoverished as soon as I could no longer welcome it by smoking...

"To stick to my decision to stop...I reduced tobacco to being only itself: a leaf that burns. I cut its symbolic links with the

Jean-Paul Sartre deconstructed smoking but couldn't quit it.

world...My regret was reduced to insignificance: I hated not having the odor of smoke, the warmth of the little heater between my fingers. But my regret was disarmed and quite bearable."

Except he couldn't think without tobacco. The foremost philosopher of the 20th Century resumed smoking; for the next four decades he was rarely seen without a cigarette burning. Even after he suffered what was probably a stroke, he defied doctor's orders and went back to his beloved cigarettes.

Shortly before he died, a reporter asked him: What is the most important thing in your life? Sartre replied, "Everything. Living. Smoking."

John Quincy Adams, the sixth U.S. President, made cigars so respectable among proper Bostonians that city fathers created a smoking section of the Common.

PRESIDENTIAL REFLECTION ONE

John Quincy Adams was seventy-eight when he gave up his prized Havana cigars, sixteen years after retiring from the Oval Office but not politics. Three months into abstinence, he was jubilant, writing in 1845 to a friend, "I have often wished" that everyone "afflicted with this artificial passion could force it upon himself to try for three months the experiment I made; it would turn every acre of tobacco into a wheat field and add five years to the average human life."

Adams died three years after penning those

words, while delivering a speech in the U.S. House of Representatives, where he still served.

PRESIDENTIAL REFLECTION TWO

Hey, hey LBJ, how many did cigs you smoke today? Lyndon Johnson was a cigarette demon—he smoked three packs a day—until he had his first heart attack in 1955, a month shy of his forty-seventh birthday. Nearly a decade later, while in the White House, the president acknowledged, "I've missed it every day. But I haven't gone back and I'm glad I haven't."

Johnson was felled by another massive heart attack when he was sixty-four.

AMEN!

One day in 1880-something, Brother Pentecost, a visiting Boston Baptist, was invited to address parishioners in London's Metropolitan Tabernacle, presided over by the Reverend Spurgeon.

The guest preacher confessed to a stormy season, seeking God's guidance on how he could show his complete devotion. Brother Pentecost boomed: "What do you suppose the Lord required of me but the one thing I liked exceedingly? The best cigar that could be bought."

The Bostonian told the English congregation that he had tried to silence the voice of conscience within him. Brother Pentecost knew how rough it is to give up smoking; his ministerial brethren had insisted he relinquish his hard habit when he had been twenty-one. He told his listeners he'd spent four years wanting to smoke every cigar he smelled.

Brother Pentecost concluded: the moment was at hand. Then and there, to his host's consternation, he offered his cigar box to the Lord, renouncing his pleasure forever.

Resuming the pulpit, Spurgeon revealed his smokier soul. "I intend to smoke a good cigar to the glory of God before I go to bed tonight," the minister declaimed. "If anyone can show me in the Bible the commandment, 'Thou shalt not smoke,' I am ready to keep it. But I haven't found it yet."

THE CIGARETTE MURDER (QUITTING THE HARD WAY)

Two-pack-a-day man, Eddie Casimir, thirty-six, strode into the deli across the street from his Manhattan apartment early on the morning of November 30, 1986. He'd just come home from a heavy date when he realized he needed a shot of Dunhill Red.

A new midnight clerk was on duty as Casimir stood on line at the busy counter. When Casimir's turn came, he asked for his regular double dose and handed the grocer a $10 bill. Just then the customer behind him cracked a joke and Casimir laughed.

The deli-man didn't hear anything funny—-he thought Eddie Casimir was laughing at him, but Casimir told police he was just waiting for his change. The clerk, a Korean, stepped out and grabbed Casimir by the lapels. The customer with the funny bone disappeared in a flash.

So here's Casimir, adrenaline replacing the nicotine which should be in his blood. And here's the clerk "disrespecting me, pushing me, acting like he hates my [Haitian] face," Casimir said five years later. When the altercation ended, the clerk was dead on the floor and Casimir was holding a bloody knife.

Casimir called the cops on himself. And when he got to Rikers Island, the city jail, he gave up smoking because prison's "no place to be needing favors."

He thought the law should go easy on him "since it was a first offense." And it's dangerous to provoke a man who needs a nicotine fix.

The cigarette killer was released after serving four years of his twelve-year sentence. And he still didn't smoke.

'We are not amused.'

Queen Victoria

Never-Evers
NOT ONCE DID TOBACCO STAIN THEIR REPUTATIONS

NERO	QUEEN VICTORIA
ATTILA THE HUN	CARRY NATION
LADY MACBETH	ADOLPH HITLER
HENRY VIII	LIZZIE BORDEN

'Do thou lust after that tawny weed, tobacco?'

Ben Jonson, 1595

Tobacco Fields Forever

lthough the English dubbed it the Indian *weed*, some tribes cultivated tobacco as they did corn, placing a fish head under each plant. Tobacco, some say, grows in sandy dirt that won't support much else, but it never grows easily. Tobacco is the most labor-intensive crop per acre but—-if the work is done right and nature cooperates—-the most profitable.

Nicholas Monardes, observing tobacco being grown four hundred years ago, didn't buy the poor-dirt theory. "Nicotiane," he wrote, "doth require a fat grounde, finely digged, the southern sonne, and to be planted by a wall."

He was describing cigar tobacco, which is now cultivated in Cuba, Central America, the Canary Islands and pockets of the

United States, Asia and Africa.

"For to sowe it," Monardes continued, "there must be made a hole in the ground with your finger, as deep as your finger can reach, then cast into that hole forty or fifty graines of the said seed."

Tobacco, botanical kin to petunias, is a golden plant, two to four feet tall, hosting brilliant flowers when nature has its way. But virgins are scarce today; deflowering and sucker-control are routine. The aroma holds true. A trace of sweet pipe smoke best mimics the leaf's natural, mellow fragrance. The seed is soft and tiny; an ounce breeds six to eight acres worth of plants.

Leaves from the middle to top part of the stem are the choicest part of the plant. Stalks used to be trashed. But now they're ground up by Big Tobacco for cigarette filler.

In North America, colonial planting took hold after Pocahontas' husband, John Rolfe, introduced a Caribbean strain into the Tidewater settlement at Jamestown.

Nicotiana rustica, the indigenous tobacco of North America is harsher than the *nicotiana tabacum*, native to the Caribbean, Central and South America. Rolfe had enjoyed southern *tabacum* after he was shipwrecked off a balmy island. Not expecting rescue, he'd smoked to his heart's content. Then he was fetched up and brought to Jamestown, where he fell for Pocahontas. He would have been a happy man, except that the local weed scratched his throat. After his imported *tabacum* was found to root and flourish in Tidewater soil, Rolfe and his family took the new, improved tobacco to England.

Cynics tell this story another way. They agree that Rolfe found *rustica* bitter, but say that his marriage to Pocahontas

Pocahontas, dressed for success in England, to sell tobacco.

was a publicity stunt brokered by Jamestown bosses eager for a quick fix to Indian/settler enmity. Pocahontas was already married to a young brave when the deal was cut. The colonists overlooked this union on the ground it was unChristian; I don't know what the Indians' excuse was. Anyway, Pocahontas, baptized Rebecca, wed John and bore a son, Thomas. The exotic family, and the exotic tobacco that accompanied them, were hits on the London social scene.

Rolfe created the most successful agricultural product of the last half-millennium. Thomas grew up and used some of his tobacco inheritance to build a fine brick house across the river from Jamestown on land left to him by his grandfather, Chief Powhatan.

The Rolfe-Warren House in Surrey, Va., built by Pocahontas' son, Thomas.

Virginia tobacco acreage increased quickly. In 1617, Virginia shipped twenty thousand pounds of tobacco to England; in 1628, a half-million. A tobacco quota, of a sort, was instituted by growers in cahoots with traders in 1639. There was such a glut of Tidewater tobacco that year that planters, fearful of steep price decline, burned half the crop. They sent to market one and a half million pounds, the same amount offered nine years before.

As world demand for tobacco jumped up. Chesapeake shores were pounded by likely and unlikely adventurers eager for a piece of the action.

The new, overnight planters were a

rough breed. The slang for their product was "sotweed." The tobacco middleman was called a "factor." One Ebenezer Cook, a self-described "gent," produced, in 1708, a ribald epic, *The Sotweed Factor or Voyage to Mary Land, A Satyr* [satire], picturing would-be planter society. He describes his landing:

> I put myself and Goods ashore:
> Where soon repaired a numerous Crew,
> In Shirts and Drawers of Scotch-cloth Blue
> With neither Stockings, Hat nor Shoe.
> These Sot-weed Planters Crowd the Shore.

Cook happens on a planter who invites him to his home:

> Whether you come from Gaol or College,
> You're welcome to my certain Knowledge
> And, if you please, all night to stay.

For supper they have cornbread, hominy grits and bacon in wooden dishes, a cask of cider for drink. After a riotous night, during which Cook's breeches, shoes and powdered wig are stolen as he sleeps, they breakfast on bear-cub meat, a gourmet item for this bear "never did but chestnuts eat."

Our traveler meets naught but scoundrels, prostitutes and wild wolves. He ships home with vengeance on his tongue:

> May Wrath Divine Lay Those Regions Waste
> Where No Man's Faithful nor a Woman Chaste.

Cook backtracks in a note: "The author does not intend by this any of the English Gentlemen resident there."

Gentlemen? There are none in the saga. But indeed, some men to manners born, and others with manners bought, amassed fortunes in the reaping and selling of tobacco. The work was unrelenting but those with high enough stakes purchased slaves to do the worst of it. In 1750, Tidewater planters shipped seventy-two million pounds of tobacco to Britain. The great James River tobacco-estate houses, such as Berkeley, overshadowed Rolfe's. They were America's first trophy homes.

Mount Vernon and Monticello stand on tobacco money. Their respective builders, George Washington and Thomas Jefferson, were tobacco planters. Both raised table food, too. Tobacco, however, was the cash crop. James Madison was also tobacco rich.

Tobacco was the resource that made North America worth claiming. It financed the intellectual capital that won the original thirteen states their freedom. Hunger for more acreage inspired the Louisiana Purchase and other U.S. westward moves. Tobacco is *why* Jefferson wanted the west won. But growing the stuff has never been easy.

In 1760, Washington wrote to a friend of being on the losing side of nature's ceaseless war on tobacco growers: "Rain for near four weeks has given a sad turn to our expectations…a good deal of tobacco [has] drowned, and the rest is spotting very fast."

Even Jefferson was sometimes discouraged. In his 1782 *Notes on the State of Virginia*, he urged tobacco planters to abandon the pursuit, and turn their attention to wheat. This, not because he was a recruit to anti-smoking (or anti-snuffing) but because growing tobacco was so much work. He wrote, "It is a culture productive of infinite wretchedness. Those employed in it are in a continued state of exertion beyond the powers of nature to support."

His advice was ignored: Jefferson sought more tobacco-friendly land. Tobacco reigned in the South before cotton and outlasted it. An 1890s jingle goes:

> Both smoking and chewing tobacco were eventually named after tobacco planter George Washington.

Cotton was once king
    And produced Carolina's cracker.
    But now we have a better thing—
    The glorious Bright tobacco.

Tidewater tobacco soil played out long ago. But tobacco continues to thrive in the Piedmont, settled by Anglo growers in the mid 1700s. The leaf also flourishes beyond Virginia and North Carolina. The U.S. "tobacco sack" today runs south from Massachusetts to northern Florida, and west from about a hundred miles inside the Atlantic coast to Missouri. It's a sack with plenty of large holes. For instance, only a small area of Connecticut is devoted to tobacco but many consider its leaf the world's best cigar wrap.

The major tobacco states are the "Bright" producers: the Carolinas, Virginia and Georgia, and the "Burley" mainstays: Kentucky and Tennessee. The main difference between Bright and Burley is in after-harvest treatment. Bright is flue-cured (charcoal smoked); Burley is air-dried and emerges lighter in color. (Cigar tobacco is also air-dried.)

Today, China grows more tobacco than any other country, almost all of it for domestic use. One brand, Panda, gained some recognition abroad because Deng Xiaoping smoked fifty of them a day, for most of his ninety-two years.

China aside, the tobacco market is international, with major U.S. based companies buying huge quantities from nations where tobacco is grown cheaply, such as Zimbabwe and Brazil.

American-grown tobacco holds nine percent of the world market because of its superior quality. Most American and European smokers depend on the flavor of "Virginia tobacco" in their cigarettes.

U.S. Tobacco growing is not mega-agriculture. The biggest farms, in the deep-South flatlands, are rarely more than a few hundred acres. That makes them large enough to host certain machinery that is useless on hilly, small farms. But even "big" growers are subject to government control of tobacco acreage.

The system is a complicated one.

The "string and horse" drying rack is a museum-piece some of today's growers know well.

But essentially, every tobacco farm comes with a grandfathered-in allotment dictating the maximum number of acres that may be planted with tobacco. A planter can buy additional tobacco acres only if an allotment within the same county comes up for sale.

What happens next is that Uncle Sam takes orders. Not from the growers, but from Big Tobacco who predict their purchases. Uncle turns around and tells the growers what percent of their allotment they may plant in the spring.

When the season is done, unsold tobacco is bought by tobacco growers' cooperatives. The growers finance this with money borrowed from the government, which they must return with interest. The co-ops sell the "surplus" tobacco, as they can.

Pittsylvania is the biggest tobacco-growing county in Virginia. The civic pride of its county seat, Danville, is wrapped in the memory of having been "the last capital of the Confederacy," the town to which Jefferson Davis fled when Union troops marched on Richmond. Tobacco, the confederacy, the pretty old homes which dot both the town and surrounding countryside, boil down to one fighting word: Heritage.

If the government makes a new tobacco-growing rule—-and there seem several hundred already—-it's an assault on Heritage. If there's a new wrinkle in anti-smoking crusades, it's an assault on Heritage.

Some insults have been swallowed, sort of. Restaurants now have smoking and non-smoking sections, but the separation is not always strictly enforced. The impression of one Danville resident is that it would do more for the health of the populace if lard were banned. But deep-fried foods and pies are part of Heritage, too.

The county shouldering Pittsylvania to the east is Halifax. Some Halifax tobacco goes to Danville for auction. The rest goes to nearby South Boston. The people of Halifax County also believe in Heritage.

It's hard to meet a Halifax County tobacco grower, or one from Pittsylvania, whose people have not been in tobacco as many generations as he or she can count back. It's also hard to meet a grower who doesn't scoff at the general notion "that tobacco is bad for you." Many in the fifty-plus generation have given up smoking, but never on principle. When a person stops, it's only because he or she is under direct doctor's orders, orders that apply particularly, but not generally.

Growers and their families live comfortably, not extravagantly. They are on easy terms with rooms Martha Stewart and Laura Ashley jumped through hoops to re-create. Antique tins rest with pictures of family on tabletops against faded floral wallpaper.

Stringing leaf was hands-on in Halifax County, Va., in the 1970s.

Larry McPeters, a veteran Halifax County agricultural extension officer, is the son of tobacco growers, and his heart is still in those fields. It's McPeters' job to tell the growers how much tobacco they can plant each year, and explain other regs. His office is government-issue, except for a miniature "Pride in Tobacco" license plate that peeks out from a stack of brochures.

McPeters reckons the average tobacco allotment in his district is twenty-seven acres. He scoffs at the idea that just about any dirt in the correct temperate zone can support tobacco. The soil in the Piedmont acres is just the soil that tobacco needs: the right "balance of macro-elements—nitrogen, phosphorus, potassium, calcium, magnesium—and micro-elements, too." Macro or micro, if the dirt is bit short on any of them, Peters knows how to dish them back in.

Halifax County boasted thirteen thousand tobacco acres in 1971; the count is now down to ten thousand. The prime attrition factor, in McPeter's view, is how "labor-intensive" tobacco growing is. "The people here work hard," he says, detailing each tobacco task, several of which are accomplished by hand.

McPeters encourages farmers to diversify, get into strawberries, snapbeans, or ostriches. His rationale: "Growing tobacco is almost an all-year job, but the auction houses are open only thirteen weeks, which means all your money comes at once."

National stats indicate that two-thirds

of younger tobacco farmers fear for their future. More think about diversifying than do it. And the doing isn't always a kick. One Halifax farmer who added melons and pumpkins told me, "They were easy to grow but impossible to sell. The supermarkets truck in produce from hundreds of miles away. They've got their contracts, their suppliers. My stuff rotted on the vine."

Another grower scoffed, "There's no money in broccoli."

There's cash in tobacco, but usually with a catch. Nineteen ninety-six was a banner year for the Heritage farmers of Virginia because the Zimbabwe crop was wrecked by drought. It wasn't an easy year, though. It started with a four percent allotment cut. At the beginning of the harvest, Hurricane Fran hit Virginia tobacco fields, destroying some leaf and damaging most of the rest.

But Big Tobacco so badly needed what Halifax had to sell, it paid top-grade price, even for the damaged goods. The price was $1.92 per pound. Back in 1978, another high-flying year, the price was $2.05.

Tobacco sells at auction, from the floor of brick warehouses where piles of graded tobacco lie in lines. On one side of each tobacco aisle is the auctioneer, rattling off prices, and a tagger whizzing along beside him. Opposite, and moving just as quickly, are buyers who indicate intentions with blinks of the eye. Each sale flies faster than reading this sentence.

The last auctions for flue-cured tobacco are in early December, closing out a season that began with planting of the first seedlings in February. In that interval, a dozen-odd diseases with such evocative names as Mosaic Virus, may lie in wait for the plants. As many as forty different weeds, from goosegrass to yellow nutsedge, may also intrude. Wild oats, no matter who scattered them, are feared. Up to twenty types of insects also gear up to attack. The hornworm is a particularly nasty tobacco foe.

The lovely ladybug, however, is a Heritage friend.

There are lines of defense against each possible attack. McPeters is a strategy consultant to the planter-generals.

In March, tobacco fields are broken. As spring deepens, fields are listed (beds prepared, pesticides applied) and transplanting begins. Lay-by, or last cultivation, takes place in June. The plants are topped in July. Leaf pulling or priming (picking, in civilian-speak) begins then too, followed by curing.

By mid-October, when the last of the tobacco is in the barn, the selling season is well under way. December and January are for reading agricultural handbooks, planning, ordering, repairing and paying off debts.

It's not an enviable lifestyle, but to those carrying on their Heritage, it's endurable except for the attacks. Blight and insect hordes are one thing; goody-two-shoes, in or out of government, are another. Heritage blood boils at the federal tax of forty cents per cigarette pack, accompanied by threats that it could go as high as $2. Banning smoking on most military premises made planters howl, and not in laughter. The Clinton administration's insistence on Food and Drug Adminstration (FDA) oversight of tobacco produced passionate reaction.

McPeters, normally an easygoing man, puts it this way. "I look at an industry that costs the government nothing, that gains it billions in taxes. Tobacco is our lifeblood in this county, and Halifax is just one of many counties where that's true. It's our heritage, it built our schools and sent our kids to college. But here's a big boy in Washington who wants to bury tobacco. He wants to bury a piece of my soul."

And McPeters, a former high school

TOBACCO ENEMY #3

This satanic moth acquired its ill-gotten wings after dining on luscious tobacco leaves as the dreaded Tobacco Hornworm.

athlete, has never smoked.

Heritage is high on the minds of Halifax tobacco growers, Harold and Belle Boswell, and their friend Lloyd Alderson, seventy-six. Alderson and his son have a farm of fifty-five tobacco acres (and one hundred of wheat) in adjacent Pittsylvania County. The Boswells are some two decades younger than Alderson, and their spread is considerably smaller. Each could talk tobacco facts all day—-if there was the time.

"Green is an ugly word," grimaces Alderson, explaining smoke-curing. "After tobacco's cured, it's not supposed to be green," he says, offering a leaf of the right golden-brown color for my inspection.

Harold Boswell shows me the generations of planting implements he has used, some

Harold Boswell poses with seed sower (top). Older implements (left) are only a few inches long. Tobacco barn is on the right.

not much advanced from Monardes' specification of a long, sturdy finger. He displays them on his front lawn.

Across the "Virginia Scenic Byway" that passes his house is the old Brooklyn Tobacco Factory, which once made chew, but which was abandoned decades ago.

The Boswells and Alderson tend cigarette-quality Bright, which brings better prices. The Boswells are as up-to-date as tobacco growers can be in Piedmont rolling country. Nineteen-ninety-six profits, and then some, went into a new greenhouse where seedlings are nurtured in water trays.

"Most of what we have goes right back into tobacco," Belle

Boswell explains. "Our children grew up understanding that they couldn't get new bikes for Christmas because we needed a new field."

Alderson's memory is longer still. As a boy, he recalls, he would sleep on the tobacco piled on the mule-driven wagon his father drove to market in Danville.

One local grower still uses a mule for planting his couple of acres. The consensus is that he probably earns more per acre than do up-to-date growers. But, of course, the Boswells are proud of their updated operations, although their barns are vintage.

Belle Boswell sums it up: "The best part is the pride that flows through all tobacco growers, the sense of accomplishment, growing a product that has been grown through generations. It's a family business, and we were taught it was honorable to grow tobacco. It's humiliating, knowing that some people think that our product is bad. My husband's hat says, 'Pride in

Auction

Tobacco,' and that says it all."

The "pride in tobacco" theme comes courtesy of customer R.J. Reynolds, headquartered in Winston-Salem, N.C., less than two hours away.

Big Tobacco appears the grower's ally: it purchases the planter's product, offers some know-how, fights tobacco curbs. But the smart grower is also wary of corporate pals: Big Tobacco buys less and less home grown, at prices that have stayed more or less the same for twenty years. But name-brand cigarette prices have soared, way beyond the premiums necessitated by increased taxes.

When the Boswells eat out, they dine in the smoking sections of restaurants, although both have given up smoking.

"I've heard young people say that drugs and alcohol are better for you than tobacco," fumes Alderson. "Next thing, the government will want to ban meat. I hear that's bad for you, too."

Government also comes under fire for a host of regulations dictating pesticide use and the working/living conditions that growers must assure the seasonal Mexican laborers, now essential to a tobacco farm. Some planters' objections seem reasonable, others short-sighted or even mean-spirited. But the fire underlying them is fueled not only by disdain for red-tape snarls and mandated siestas, but also by nostalgia for tobacco farm work as it was only twenty or thirty years ago.

Family members of all ages worked side by side then, planting, priming, hauling, stringing and curing. They accomplished this with the seasonal help of neighbors, white and black, who were willing to share the long hours, the yellow-nicotine stains on their hands ("an occasional feller has allergic skin," Alderson muses, "but it's never irritated me none") and the broil of the summer sun. The summer help, most of it young, was paid a very modest wage and often shared meals with the grower's family.

Today's teens, the growers and McPeters say, would rather stay in bed and watch MTV.

Some growers' kids have turned their backs on the crop that raised them. Charlotte Stillman, seventy-seven, who was widowed last year, was the daughter of growers and the wife of one. "Tee," as she is known, is as gracious a southern lady as ever benefitted from Virginia's

Maryland on Our Mind

Lord Baltimore's charter specified that tobacco be planted, and Charles County has long been a leading producer. So it figures that when a beauty is crowned at the county fair in La Plata, Md., she is known as Queen Nicotina. The 1996/97 reigning monarch was Laina Baumann, 17, old enough to be lovely but too young to smoke.

loam. Heritage has not made her rich. In the ante-bellum house built by her great grandfather, she smokes generics to keep her expenses down.

Tee and her husband, John, raised six children on their fifteen acres. After John contracted emphysema, none was interested in working his fields. The tobacco acres, now down to seven, were rented out.

John Stillman stopped smoking when he got sick, but Tee says, "he loved the aroma of tobacco until the day he died.

"I stopped for a week when he did, but the whole world got so bleak—-nothing was fun. I got so mean, I went back to smoking. I was reading that if it hasn't killed me yet, it probably won't."

Tee Stillman remembers, "John had some real good friends across the road who'd help him plow and bring the tobacco in, so we never much depended on hired help. John and the men would put the leaves on a long table in the barn. My boys and I would hand it to the stringer. She was a young black girl who specialized. John must have paid her.

"When I was a girl, though, my job was helping my mother feed the hands. We'd be up before dawn to churn the buttermilk, and for the midday meal they'd get the buttermilk, cornbread and vegetables from the garden we'd cooked up.

"It was all much easier by the time I was the one in charge. But my children, they couldn't wait to leave here."

None of her sons smokes, although two daughters do. The third daughter hates tobacco smoke so much she won't let anyone, her mother included, smoke in her house. Charlotte Stillman laughs when she tells that. Then she shakes her head: "Times change, but I don't know where the South would be without tobacco."

Charlotte Stillman, a grower's daughter and a grower's wife, who once saved the harvest from a hurricane with only the help of her two young children, has no regrets.

'When they talk'd of their Raphaels, Correggios and stuff,
He shifted his trumpet And took only snuff.'

Oliver Goldsmith
Retaliation, 1774

The Right Snuff

When you stick your nose in tobacco history, you come up sneezing.

Tobacco has been known in the west for five hundred years; for three hundred of those, people of fashion have put it up their noses. The result can only be met with: God bless you.

Inhalant tobacco is named snuff. The French, who turned tobacco snorting into royal play in the 16th century, called snuff *tabac en poudre*, or *rapee*, the latter term still current on certain labels. The French called snuff users *priseurs*.

You don't meet many *priseurs* any more, although there are still enough jaded noses in London for Harrods to stock several snuff varieties, including Hedges Carnation. One Covent Garden haven bills itself as a snuff parlour, occasionally offering Keen Scental to wine-sipping patrons. Just how haughty today's partaking English schnozzes are, is open to question.

In the mid-19th-century, snuff slid down the American social scale where it remains today.

The stuff that is snuff is always powdered.

The snuff of aristos is inserted in the nasal passages. The immediate consequence is a jolt, followed by sneezing.

There is also dipping snuff, a moister concoction, which goes straight to the mouth, to be mashed between the gum and lip. The immediate consequence is a jolt, followed by spitting.

Plain old chewing tobacco is

sometimes mistakenly referred to as snuff. But that's the wrong stuff, although its aftermath is also spitting. Chew is another chapter.

Snuff is often flavored. Native South American chefs sometimes dashed it with cocaine powder, which may be why its rep for clearing the head loomed so large. But even snorted straight, the tobacco ground from prime leaves is potent.

Over the centuries, hundreds of seasonings have been added to snuff. Here's a 17th century recipe for making it at home.

- Spread and dry the best sort of tobacco leaf in the sun.
- Strip the leaf from the stalk.
- Put it into a mortar and beat it into a powder.
- Push it through several sieves til it runs fine.
- Take a vessel bigger than that which will hold the tobacco dust and line it with a napkin of strong cloth.
- Pour rose water, or any sweet water you like, over it.
- Make it as a paste.
- Dry it again.
- Repeat until it is as you please.

Snuff had not been long in vogue before dozens of ethereal mixtures were commercially offered. The pulverized tobacco was not only tarted up for aroma, it was dyed; red and yellow were favored hues. One tobacco antiquarian catalogued no fewer than a hundred flavors sold in England in 1728. Available today are tobacco powders blended with cinnamon, fruit,

This colonial gent, snuffbox in hand, has greeted customers of Demuth's Tobacco Shop in Lancaster, Pa., since 1770. Gen. Edward Hand, on George Washington's staff, bought the right snuff there.

DEMUTH FOUNDATION

THE SMOKING LIFE **103**

mint and other herbal essences.

French sniffers were soon joined by enthusiasts in Ireland, Scotland and Italy. The Russians were enthusiastic about snuff, too, until a czar took an ignoble stand against it, decreeing in 1634 that users would be deprived of their nostrils.

The English upper classes, supplied by Scottish merchants, converted to snuff with the restoration of the monarchy in 1650. The kilted Highlander, thumb and forefinger pinched together, became the sign of the snuff merchant. Satirist Henry Neville commented that same year, "She that with pure tobacco will not prime her nose, can be no Lady of the time."

Lesser mortals got in on the Big Sneeze during the plague, when snuff acquired undeserved cachet as a prophylactic. Almost everyone seemed to get off on snuff. As the Moliere character, Sganarelle, remarked in 1665, "Not only does it exhilarate and clarify the human brain, it instructs mortals in the ways of virtue and influences one to go on being pleasant."

Hot snuff. So hot, it fueled a century of religious controversy. Although Liz I early on banned snuff-taking during Anglican services, her rules did not apply to the Roman church.

Foremost in the annals of embarrassing ecclesiastical moments is an anecdote recounted by A. Vitagliani in his 1650 *D'Abusu Tabaci*. He tells of a Neapolitan priest who treated himself to a pinch of snuff immediately after communion. Still at the altar, he began to sneeze so violently that he uncommuned, as it were.

Twelve-Step Snuff Program

From "A Letter to a Young Mademoiselle" penned by Charles Perrault in 1750.

1. Take the snuffbox in the left hand.
2. Tap on the snuffbox.
3. Open the snuffbox.
4. Offer the snuffbox to the company.
5. Take back the snuffbox.
6. Gather the snuff together by tapping one side of the box.
7. Take a pinch in the right hand.
8. Hold the snuff in the fingers for a time.
9. Carry the tobacco to the nose.
10. Snuff with gusto in both nostrils without making a grimace.
11. Sneeze, cough, spit.
12. Close the snuffbox.

It was after the host was spit up in Naples, that Urban VIII banned snuff taking by priests while celebrating mass. The pope tacked certain categories of laity onto his order, a rule Benedict XIII rescinded to keep up church attendance.

Seventeenth-century theologians debated whether snuff is forbidden nourishment during Lent. (Rabbis have had to contend with whether smoking is legal on fast days, or the Sabbath which is supposed to be labor free. The Jewish dilemma is twofold: Is tobacco a food? Does setting it afire constitute work? I was unable to ascertain if a recent recording, *Cigarettes/Happy Jews*, by the Klezmer Conservatory Band on the Yiddishe Renaissance label, has bearing on these issues.) Some 1600s thinkers stuck to the view that tobacco was tainted by its pagan origin, but others advanced the idea that tobacco, if not next to

Some Londoners still consider tobacco snorting an enlightening, if not religious, experience.

godliness, was at least God's handmaiden in tamping down hungers, lust included. In 1669, Benedetto Stella concluded: "The natural cause of lust is heat and humidity. When dried out through the use of tobacco, libidinous urges are not felt so powerfully."

Religious orders took to snuff in a big way.

Rhymes, if not poetry, have been dedicated to the joy of snuff. This one was composed by an anonymous Italian in the thrall of a Scottish brand:

What a moment! What a doubt
All my nose, inside and out.
All my thrilling, tickling caustic
Wants to sneeze and cannot do it.

THE SMOKING LIFE **105**

What shall help me? Oh, good heaven!
Ah yes, Hardem's thirty-seven.
Hang it, I shall sneeze til spring.
Snuff's a most delicious thing.

Among history's celebrated snuffers are Rousseau, Pope, Swift, Congreve, Addison, Gibbon and Metternich. The supreme French diplomat, Talleyrand, once listed the ancillary charms of snuff use:

- Serves as a pretext for delaying a reply.
- Sanctions looking away from questioner.
- Keeps nervous hands busy.
- Keeps face from displaying feeling.

Although it insulted fashion to speak of it, the journey to the perfect sneeze via tobacco powder had its health burden. Interesting germs sprayed every gentrified occasion. And in 1791, a noted London physician, John Hill, diagnosed several cases of nasal cancer in snuff-takers. Today's charming tins state: "Tobacco seriously damages health."

F. W. Fairholt, a 19th-century journalist who grew up in a tobacco warehouse, described the snuffing styles of the early 1800s. "Some do it by fits and starts," he wrote. "These are epigrammatic snuff-takers who come to the point as fast as possible and to whom pungency is everything.

"Others are all urbanity and polished demeanor. They value the style as much as the sensation, and offer the box around out of dignity.

"Some take snuff irritably, others bashfully, others in a manner as dry as snuff itself, others with a luxuriance of gesture and a lavishness of supply, that announces a moister article and sheds its superfluous honors over neck-cloth and coat."

Contemporary made-in-America "Scotch" snuff.

For the moneyed classes, the pleasure of snuff was enhanced by its many accoutrements. Eurotrash did not invent all of them. The well-equipped Amazon shaman included among his possessions a small wooden plate and mortar for snuff manufacture, plus a peccary brush for sweeping the powder into the bone-tube nasal applicator. But rich Europeans gussied up snuff implements: carving, painting and bejeweling snuffboxes and equipping them with tiny graters and masterfully-turned gold and silver spoons.

Some snuff notables endeavored to keep their habit low key. Samuel Johnson, who reluctantly took up snuff, kept his stash in a waistcoat pocket. Robert Burn's horn snuffbox, with a silver plate engraved with his initials, sold at auction for next to nothing after his death, emerging only much later as a museum piece.

The Scottish poet's snuff box was humble, compared to many wrought of ivory or gold, enameled or of hand-painted Meissen. World-class gems were studded in tobacco-powder con-

An elegant snuffbox.

tainers which did double duty as music boxes. A gentleman might chain his snuffbox to his waistcoat; a lady might wear it as neck, wrist or waist ornament.

On occasion, the snuffbox was the high stake in court gambling games, such as Basset Table. Lady Mary Wortley Montagu versified such a wager in 1716:

> This Snuff Box—on the hinge see
> Brilliants Shine.
> This Snuff Box will I stake,
> The Prize is mine.

The snuffbox was the perfect gift. Marie Antoinette received fifty as wedding presents. The expense ledger for the 1820 coronation of George IV includes £8,205 spent on

THE SMOKING LIFE

From Eternity to Here

Royal Victoria Snuff

A renaissance of upscale, inhalant snuff wafted through select Madison Avenue boutiques in 1966. It happened this way. The Lyons, a young New York couple, in need of a Victorian sofa, answered a newspaper ad placed by the Smiths, a young New York couple, with a Victorian sofa for sale.

The Lyons liked the sofa and the Smiths. The four, in a Victorian mood, adjourned to Chinatown to celebrate their new friendship. Browsing in a Chinese herbal shop, they came across a fifty-cent canister of plain tobacco snuff. Almost instantly, four great minds were seized with a single bad idea: they would go in the snuff selling biz.

Not any old snuff, but the right snuff, dubbed Royal Victoria and flavored with cocoa or vanilla, just as in days of old. The hundred pounds of powdered tobacco they ordered came to rest on the Smiths' dining-room table—because they had one. The four enthusiastically set about creating snuff flavors. Elegant containers were devised; a cunning leaflet printed to accompany each one. Nan Lyons was named Director of Sales.

How could the firm of Smith and Lyons lose? It was a unique product in its time and place, and the markup—each vial would retail for $12—was astounding.

Alas—difficulties beset the fledgling manufacturers. The blending proved unexpectedly hazardous. The four chemists sneezed nonstop; asthma forced one to retire from this phase of the business.

The sales force under Nan Lyons consisted of Nan Lyons, who met a cooler reception on Madison Avenue than she'd expected. "They hated me," she recalls. She persisted, and certain select shops began offering Smith and Lyons for sale.

Some even sold. But not enough. A hundred pounds of sniffing tobacco, cut with other powders, equaled just too much Royal Victoria Snuff.

---

snuffboxes for foreign ministers.

A nose tickler named George IV is sold today in Britain. Other English inhalants come with tweedy labels: Jockey Club, Crumble, Jock's Choice.

In the United States, some nasal snuff travels under Scottish stickers, powdered up by Appalachian descendants of long-ago emigrants from the tartan isles. Dr. Rumney's Mentholyptus Snuff is so powerful it set me sneezing as soon as I unsealed the tin.

As Cro-Magnons once overlapped Neanderthals, sniffers and dippers have long coexisted.

Dip snuff is four-alarm tobacco. Forget how it's been candied, you have to love the taste of tobacco to sop up dip. The nicotine hit is major. It wakes you up completely. Ninety per cent of the snuff sold in America today is of the dip variety.

Dippers with senses of propriety also had paraphernalia. Knives, razors or graters were essential when American snuff tobacco was fashioned in ropes or twists that required cognoscenti to shave off portions for use. Even after snuff was packaged in convenience form, a stick and a spit cup came in handy. Although the sterling or prized porcelain spit cup was not unknown, most were (and are) of more modest origin.

A Virginia lady of my acquaintance recalls her grandmother's dipping ways. As a widow, her grandmother "lived about," meaning she moved among the houses of her grown children, bringing her long cotton dresses, black stockings and snuff cup with her. Her granddaughter remembers, "She kept her snuff—a brown powder, drier than what's around today—in one apron pocket, and a good twig in another. She'd dip with the twig, and she'd use her special cup for spitting. It was very discreet. Dipping was done only in your home. You didn't go visiting and do that if you were ladylike."

A lifetime New Yorker I know is a closet dipper. His closet is his jeep. He conceals his snuff in a brown-paper bag in the glove compartment, and only dips, he says, when he is in imminent danger of falling asleep at the wheel.

In much of Western Europe, dip is thought so repulsive—and so conducive to mouth cancer—that it's banned. But taking dip or *snus* is one of the smiles of a summer night in Sweden. Among Swedish young men, dip is more popular than cigarettes. Grov, General and Goteborg's Rape are the big labels; the American brands, Skoal and Copenhagen, have small followings.

Mouth-watering snuff, a.k.a. dip.

One encounters most American dip enthusiasts in rural regions, Marlboro Country among them. Dip is popular with carousing car-racing fans, and among the caddies at the elite Maidstone Club in Easthampton, where they're not allowed to smoke. The implements of choice for most of today's dippers are the fingers. In the great outdoors, the spit cup tends to be the good earth.

Snuff said.

'In these raw mornings when I'm freezing ripe, What can compare to a Tobacco Pipe?'

Samuel Wesley, 1685

Piping Up

Pipes are the pleasure of professors and sea captains, or so many of us think. Not.

Information about the three million American purchasers of pipe tobacco only allows the conclusion that the people who put that in their pipe and smoke it are men. Old, middle-aged and young men, rich and poor men, educated and uneducated men are all equally likely to buy the coarse, long-cut tobacco offered for pipes.

Slightly more pipe tobacco is sold in the American South than in any other U.S. region, which figures because old habits last in Heritage country. The British male is another creature fixed in his ways; one of those ways for some older men is clamping down on a pipe.

The rising tide of cigar interest has produced eddies of infatuation with pipes. That makes sense because if you're looking for something to smoke that isn't a cigarette, and you want what you smoke to come with unlimited possibilities for emptying your wallet, you'd be foolish to overlook pipe potential.

Shag doesn't cost much and some pipes are bargains. But a pipe can also set you back hundreds of dollars, and you can display as many as you can afford. If you get into pipes, as opposed to simply smoking one, you can also stock up on racks and tobacco pouches, cases and humidors.

It all smells wonderful. I'm not an adept at pipes, and find the tiny but acrid kickback, well, distasteful. Yet I love the aroma and happily sit at the feet of pipe smokers. I actually invent reasons to visit people who smoke pipes.

THE METROPOLITAN MUSEUM OF ART, LAUDER COLLECTION OF AMERICAN POSTERS, GIFT OF LEONARD A. LAUDER, 1984. PHOTO BY BOBBY HANSSON

Pipes make you appear thoughtful. Even watching a pipe smoker may add points to your IQ. Frederick the Great probably considered this when he convened the *Tabaks Collegium* as a forum for debating Prussian policy. Pipes (and snuff) were *his* heritage, and he knew that tamping, lighting and drawing on a pipe all necessitate pauses in conversation, pockets of time to actually listen to what someone else is saying. Reviving the *Tabaks Collegium* is what made Frederick *great*.

Or almost great—for the *Collegium* was not an equal-opportunity smoking event. Although pipes and women, and wisdom and women, rested on strong historical foundation, the queen lit Frederick's pipe to mark the session's beginning and then withdrew.

Occasional *Collegia* were social gatherings where women were included. Eighteenth-century society thought it more genteel for a high-born European lady to sniff than to smoke, a circumstance some protested. In classy American circles, pipes for women lingered longer. At least one 19th-Century First Lady, Margaret Taylor, was known to enjoy hers in the

White House.

Pipes stayed in style for women who paid no notice to fashion and vice versa. Pipe smoking was not extraordinary for women in London slums and in the picaresque countryside on both sides of the Atlantic. But by the century's turn, pipes had been knocked out of feminine hands by trickle-down Victorianism, or replaced by naughty-girl cigarettes. The revolution was never complete; there remain a few good women who are rebels with a cause, that cause being a pipe.

The pipe could be defended on aesthetic grounds alone. I know of no utilitarian object that has been crafted in more diverse ways. Snuff box decoration is imitative of other arts, and sometimes gaudy compared to the many graceful forms of the pipe.

There are thousands of specimens of Native American pipes, each an original carved or sculpted work. Some were in use when Europeans bumped into the Americas; others, older still, have been recovered from burial mounds. Indians piped on any conceivable occasion; one Central American pictograph shows

Pipes stayed stylish for women who didn't care about style.

women smoking while bathing.

The craftsmen of the Americas often modeled their pipes on birds and animals; some stacked form on form, creating totemic works. They fashioned pipes of any amenable material known to them. Tortoise shell, silver, gourds, clay and wood are just a few of the substances used. Some Native American pipes don't look like pipes to the untutored eye; a pipe may have no obvious stem, or it may be mostly stem, flared at one end for tobacco insertion. The tube pipe was also used in close proximity to a separate bowl.

Montezuma smoked cigars, sure, but there's also an account of him selecting an elaborately-painted tube pipe from the smoking table offered by maidens at the conclusion of a three-star Aztec meal. The tobacco for that pipe was freshly ground in a fragrant rosewood bowl before presentation.

For ceremonial and social occasions, members of several Indian nations used baked-clay or reed straws to smoke from an earth bowl, a hollow in the ground in which a tobacco fire was laid. When bereft of a stem, some Native Americans dug two holes in the ground. The hole on one side was filled with lit tobacco. At the opposite hole, the hard-core pipe man lay on his stomach and puffed. This smoking method was copied—or reinvented—by enterprising nicotine lovers in Africa. Such is the art of need, which may be as impressive as the creativity behind artifact.

The 'White Man' invented the tomahawk pipe to trade for 'Red Man' goods.

114

One of the more curious pipes ever fashioned is the tomahawk pipe, manufactured in England expressly as a trade item to be offered North American Indians by merchants hankering for glossy pelts. This hybrid had an axe at one end, an alleged puffing device at the other.

The clay pipe was the most popular pipe type among Europeans on both Old and New World soil. Jamestown seems to have had its own pipemaker, Robert Cotten. As soon as tobacco hit Holland and England, the craft of pipe making took root. European pipes generally have had smaller bowls than American ones, reflective of the higher price of imported tobacco.

In the Near East and North Africa, tobacco rapidly joined or displaced other smoking herbs in the hookah. Clay pipes also won converts. Africans carved lovely specimens from wood.

In China, clay pipes hosted opium until the new challenger appeared. The choice between opium and tobacco remained open in China and elsewhere for centuries. When Sherlock Holmes refers to a "three-pipe problem," it's not clear what will fill each of those bowls.

One's chosen pipe depended as much on income as preference, and choices ranged from the humble North American corncob pipe to elaborately-carved ivories. In clay, there were many grades, culminating in fine porcelain. Some stalwarts maintained that the best clay pipe was the familiar one, as this 19th-Century ode by A.B. Van Fleet attests:

Moravian clay pipe bowl.

> There's a lot of solid comfort
> In an old clay pipe, I find
> If you're kind of out of humor,
> Or in trouble in your mind.

Many hold that the finest clay pipe ever made is the gambier. The bowl, sometimes elaborately sculpted, is affixed to a wood stem and capped with a vulcanite mouthpiece.

The Gambier was the *fourneau* (French for both "pipe bowl" and "furnace") of choice of the 19th-century poet, Arthur Rimbaud, who said he confronted the world with *"une gambier aux dents."* Rimbaud may have been an incurable romantic, but he was not a neat smoker. He confessed in a letter, "I smoke my pipe by the window and spit on the tiles."

Tobacco burns well in a gambier, experts say, but the pipe draws slowly and the bowl, as any of clay, is fragile.

The meerschaum has been titled the queen of pipes, as befits an instrument delicate and expensive. Meerschaum, in German, means "sea foam," and the material, a hydrated magnesium silicate, primarily found in Turkey, is an artist's dream. The silicate is milk-boiled, oiled, polished, waxed and glazed before the craftsman's work is done. With age and use, the meerschaum mellows from off-white to yellow into brown. The most elaborately-sculpted pipes are meerschaums.

The bowl of this Victorian meerschaum is a Sikh.

The meerschaum displays its witty artistry on the exterior of the bowl and, possibly, along part of the stem, too. From the bowl may emerge a female figurehead, a famous or comic face, a pair of stags—indeed, any form that the pipe maker or the smoker, who commissioned the meerschaum, fancied.

A hot meerschaum set on a cold surface will crack.

Certain woods were long favored for prestigious pipes. Rosewood adds scent to the smoking tobacco but becomes hot and always requires a good mouthpiece. Cherry is admired for its look, but pipes made of it are said to char extremely.

No wood was trouble free until *bruyere* or briar, became the class act last century. It's the root of the Mediterranean briar that is used for the pipe. The briar began its ascent to the top of the pipe heap after a French pipe manufacturer acquired one, while on holiday in Corsica. In fine pipes, the briar is worked so that its grain goes the same way in the bowl. Most of today's briars are married to dark Lucite stems.

Bowl size and shape, as well as stem length concern the pipe aficionado. Among the more significant styles are the long-stemmed churchwarden (elegant but its stem can snap), the short, squat bulldog and the billiard, which is the style of most briars.

Mouthpiece is a another factor. Alfred P. Dunhill of London, a maker of automobile accessories, first found favor in the smoking world by promoting amber as a pipe mouthpiece and as a cigarette holder.

Classic briar.

English pipe makers are tops in the hearts of many, but some believe briar is still best worked in France, or by a cult of Italian carvers said to have access to a superior supply of the gnarled root. Distinguished British pipe emporia tend to stock several imports, but with the claim that a particular pipe has been made to its specifications and under its supervision.

A pipe lover might want a selection of woods, a Dutch or German porcelain and a meerschaum to vary his pleasure, or just show off. Then there is the notion of loyalty to one's original pipe. The first pipe smoked may nostalgically be the best, but never the easiest. Read the lips of Anonymous:

Oh Indian weed, Tobacco bright,
(But stay, first let me get a light)
The choicest gift the world e'er saw,
(Confound this pipe! Why don't it draw?)

Thou art of plants the noblest gem
(There's something sticking in the stem)

Thy healing properties, none doubt.
(The knitting needle's got it out.)

Of human pleasures through the Crown!
(I shall be better lying down.)
Oh anodyne of mental pain.
(You won't catch me at this again.)

HOW TO SMOKE A PIPE

First, buy a good pipe. The troubles caused by bad pipes —especially those with thin bowls or too-small holes in the stem—are all nasty. You can burn your tongue, sting your tastebuds and worse. Clean the pipe before

Native Americans created astonishing variations on the pipe theme.

SMITHSONIAN INSTITUTE

you make it your friend. Seafarers used to fill the bowl with rum and let it stand for a few days. You probably shouldn't drink the dregs, just tinkle them overboard and wipe the cup with a clean cloth. Make sure the passage from bowl to stem is clear.

Next, buy tobacco that smells good. Since pipe tobacco usually comes in sealed bags, you may not be able to whiff it before purchase. So take advice, or plunk for few varieties and be your own test kitchen. Also buy long matches, lots of 'em, or a reliable lighter whose flame doesn't lick your fingers. Do your shopping before you go to sea.

Layer in the leaf; tamping down each pinch. The tobacco should lie a bit looser at the bottom of the pipe than at the middle, and be most dense at top. But don't fill your pipe all the way until you've broken it in. What you want is a nice inner carbon char and that must be built-up gradually.

Light your pipe from the middle outward, and take a few puffs to spread the flame. Since you don't inhale pipe smoke you'll have to keep your mouth open, while clamping the pipe in your teeth, so the smoke can exit. Stop puffing—-let the flame go out. Your tobacco should be crisp on top now. It's okay to leave the burnt stuff there, just smooth out the bumps. Now light up again, and go full throttle.

It takes practice before you can smoke a pipe and do anything original at the same time. When you're done, don't forget to tap out the dottle.

'I think he is an excellent host, he has travelled... he has some excellent cigars.'

The Once and Now Cigar

Cigar smoking isn't an easy-going habit, it's a lifestyle, or so the purveyors of premium smokes stress. Connoisseurship requires mastery of a complex new vocabulary of cigar differentiation, humidors correctly gauged for temperature and humidity, an elegant, efficient tool for clipping the cigar and knowledge of the technique for lighting it evenly.

The world being what it is, you also need a welcoming space in which to enjoy your cigar. Moreover, should you choose to indulge in a prestige smoke from Havana, you may need a *connection*. The personal smuggler is the personal trainer of the late '90s.

Since bigger cigars contain as much tobacco as two packs of cigarettes, it is the rare enthusiast who enjoys more than a few each day. This means it takes years of tasting to become an epicure.

In the United States, an over-the-counter cigar, in the premium category, costs anywhere from $2.50 to $50. A classy Manhattan outpost sold me a classy Robusto for under $5. Cigar bar menus usually offer sticks at $6 or $8, then climb from there. Value is in the tastebuds and pocketbook of the consumer, the content and rarity of the product, and the trappings and integrity of the seller.

You should remember, too, that smoking primo cigars suggests a range of other exacting pleasures: gourmet dining, imbibing premium spirits, golf playing and ownership of a stable of racehorses. A yacht wouldn't hurt.

On the other hand, you can pick up a five-pack of machine-mades (White Owl or

Philly Blunts, for example) for about $1.75. Fifty cord-bound cheapies can set you back as little as five dollars for the bundle—-that's a dime apiece.

Also on the market are cigars in which the filler and binder are machine bunched, but each bunch is hand wrapped.

If you live in New York or in one of several Florida cities, you can wander into a small cigar-making establishment and purchase a fresh, hand-rolled smoke for a buck or so. But do you want to do this?

I can't read your soul. I can only tell you that inexpensive cigars still outsell premiums, although it's the luxury end of the market that is exploding.

Despite the boom in cigar emporiums, cigar dinners, cigar-of-the-month clubs, cigar videos, cigar bars and, in New York, a cigar disco named Decade, there is no typical enthusiast. The education and income levels of cigar smokers are unpredictable.

It's a safe bet, however, that not many dime-cigar smokers elbow their way into Hamilton's, the Los Angeles cigar club run by George Whatshisname, who

"Cocktail" by Gerald Murphy, 1927

was once an actor. Fewer frequent the swank Grand Havana Room, poised for bi-coastal success. American cities, however, abound with cigar "clubs" that do not demand you pay dues, only your bar check.

Informal smoker hang-outs also thrive. Cigar love tends to be a city thing, favored by people between the ages of twenty-five and forty-five. The cigar's status as a city light is likely to burn brighter as yuppie identification with the power smoke intensifies.

Tomorrow's yuppies are already experimenting with choice and not-so choice brands. Prep school students show off with cigars. The nonsmoking mother of one such sixteen-year-old said, "I'd rather he didn't, but as rebellious activities go,

Power Smoker: The Babe.

this one is relatively harmless."

Cigar smoke wreaths some private offices where big deals are cut and haloes some weekend athletes in the once-great outdoors. Cigars are smoked on Aspen ski lifts by the kind of guy who make cell-phone calls at the same time. The "Tee-Gar" has been invented for the golfer who wants to rest his smoke while he takes a stroke.

The Cigar Institute of America claims there are two hundred thousand women cigar smokers. The red-hot cigar-smoking mama can be spotted at cigar bars, but she's not a ubiquitous feature, although a with-it Manhattan restaurant offered a Mother's Day cigar dinner in 1997.

The first cigar fans were probably casual about their smokes—ripping and rolling leaves from the beloved plant as they felt the urge. The cigar's outer layer, commonly called the wrapper, was either leaf or corn husk. Some tribes tended plants to ensure supply and some leaf was stored, effectively air-dried. This cured leaf may have been the preference of Cubans by the time Christopher Columbus barged in.

The Aztecs knew how to enjoy a cigar.

Columbus knew even less about cigars than he did about geography. Actually, he knew *nothing* about tobacco, an understandable shortcoming since (as far as we know) he was among the first Europeans ever to lay eyes on it. He logged, but could not identify, the "esteemed" dried leaves he was offered on October 12, 1492.

The Award for Discovering What

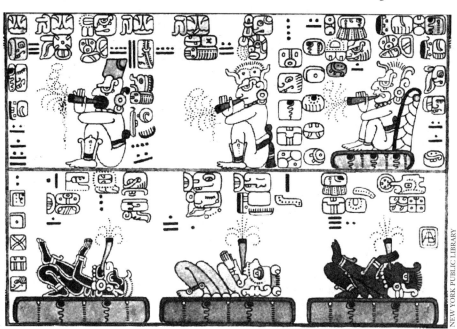

Was Already There goes to Columbus' scouts, Rodrigo de Jerez and Luis de Torres, who unmistakably described cigars to their captain after their poke around Cuba the next month. De Jerez, as we've learned, became the first white-man *aficionado*, and was jailed for his affectation by the Spanish Inquisition.

The Spanish eventually overcame their fear and loathing of the unfamiliar, setting up cigar factories on home soil. The Royal Manufacturers of Seville was chartered in 1731. The Spanish remain cigar mavens. Tabacalera, Spain's tobacco company, annually produces three hundred and fifty million cigars, and buys half of all handmade Cuban product.

The English-speaking world was slow to catch on, clinging to its "discovery" of tobacco in pipes, even as its upper echelons toyed with snuff. The tobacco tube did not hit North America in a big way until 1762 when Israel Putnam, a British Navy colonel and veteran of England's contretemps with Spain over Cuba, imported Havana cigars to the northeast colonies.

The rolled leaf came to most of Europe via Spain, but not until the early 19th-century Napoleonic Wars blew cigar smoke in all directions.

Catherine the Great of Russia is said to have invented the *Ur* band, insisting on white silk ribbons on her cigars so her fingers would not be stained yellow. Cuban cigar makers adopted bands as condoms against brand misrepresentation, not nicotine.

It was not until after the Revolution that cigars were first rolled near Pennsylvania tobacco fields, seeded from Cuban stock. Hamburg became the first Northern European city to make cigars in 1796; Bremen also became a big center. The French, Italians and Swiss started manufacture in the early 1800s. England relied on foreign goods; import tax was paid on fifteen thousand cigars in 1823; thirteen million in 1840.

Philadelphia, New York and Tampa, Fla., were hubs of American cigar making. As puffing cigars turned to popular sport, belle lettrist James Russell Lowell commented on their deteriorating quality. In 1839, he wrote dramatically to a pal, "Smoking is not what it once was.

Ulysses S. Grant, cigar poster-boy.

Weep with me, friend of my bosom, on the degeneracy of cigars! Men and cigars decline together...The filling of cigars now belies the wrapper. So with men: they have a very well-seeming outside of learning or ideas, but are not so well-filled as of yore."

Gen. Ulysses S. Grant emerged as America's cigar poster boy. Grant treated himself to twenty stogies daily. (Robert E. Lee, leader of the Confederate forces, did not smoke—-talk about losing causes.) Union supporters heaped thousands of boxes of cigars on Gen. Grant after reports of his blazing to victory reached the homefront. The ever-increasing stash followed him from campaign tent to tent and eventually to the Oval Office. The general became the occupant of Grant's tomb after succumbing, at age sixty-two, to throat cancer.

Mark Twain, né Samuel Clemens, outdid Grant. Twain once defensively remarked he "only smoked one cigar at a time." He relished thirty such times each day, relieving the intervals with a pipe. Twain, who lasted nearly three quarters of a century, outlived most of his critics.

Those critics were divided between those who couldn't stand *that* he smoked and those who couldn't stand *what* he smoked. Twain once declared, "Nearly any cigar will do me except a Havana."

He skewered the cigar snobbery of his time (which sounds a lot like the snobbery of our time). In his 1890 essay, *Concerning Tobacco*, he wrote, "People who claim to know say I smoke the worst cigars in the world...

"Now then observe what superstition, assisted by a man's reputation, can do. I was to have twelve personal friends to supper one night. One of them was as notorious for costly and elegant cigars as I was for cheap and devilish ones. I called at his house and when no one was looking borrowed a double handful of his very choicest...I removed the labels and put the cigars into a box with my favorite brand on it—-a brand which those people all knew, and which cowed them as men are cowed by an epidemic.

"They took these cigars when offered at the end of the supper, and lit them and sternly struggled with them—-in dreary silence for hilarity died when the fell brand came into view and started around—-but their fortitude held for a short time only; then they made excuses and filed out, treading on one another's heels with indecent eagerness; and in the morning when I went out to observe results the cigars all lay between the front door and the gate.

"All except one—-that one lay in the plate of the man from whom I had cabbaged the lot. One or two whiffs was all he could stand. He told me afterward that some day I would get shot for giving people that kind of cigar to smoke."

Nineteenth-century cigar men tended to prefer their self-referential pleasure to female company. Back then it was a rare gentlewoman, outside Latin America, who would tolerate smoke in her presence, never mind join a man in sucking a cigar. French novelist George Sand (née Aurore-Lucile Dupin) was an exception; she was as well-known for her smoking as for her cross-dressing. Her first live-in lover, Jules

Women served the cigar world on labels.

Sandreau, noted, "The smoke of the cigar produces the same effect as opium, in that it leads to a state of febrile exaltation, a perennial source of new pleasures."

Another exception was Kate Carrington, who set her feelings about the perfume of cigar smoke to verse.

The aversion of most ladies to smoke contributed to the popularity of men's clubs as a retreat for lunch or from life, and fostered the Victorian custom of segregating the sexes after dinner. The women retired to freshen up or chat, the men to huff and puff. Rudyard Kipling's "a woman is only a woman but a good cigar is a Smoke!" sums up the misogynist view.

The cigar on which the sun never set was the square-cut cheroot, often made then from Sumatra and Java tobacco. Cheroot is a Tamil word. The Tamils lived in Ceylon, now known as Sri Lanka. Most prized by the English was the Burma cheroot.

The pick of the lads in the barracks was not the first choice of London gents, who favored select Cuban product. "By the cigars

Ode to Cigar Smoke

What is it that comes through the deepening dusk—
Something sweeter than jasmine scent,
Sweeter than rose and violet blent,
More potent in power than orange or musk?
The scent of a good cigar.

And what does it say? Ah! That's for me
And my heart alone to know;
But my heart thrills with a sudden glow,
Tears fill my eyes and I cannot see—
From the scent of a good cigar.

—Kate A. Carrington

they smoke and the composers they love, ye shall know the texture of men's souls," John Galsworthy wrote later in *Indian Summer of a Forsyte*.

Havanas were status symbols in the United States, too. Most Americans, however, smoked cigars made closer to home. Samuel Gompers, who established the American Federation of Labor, got his start in the 1870s, organizing immigrant rollers in dark, dank sweatshops. Gompers worried about workers' health as their skin sopped up yellow sap, and he decried their meager wages and their boredom.

One workplace improvement was the hiring of a reader, a person who entertained the immigrant Cuban rollers by reading aloud classic, literary works.

In 1900, ten thousand cigar factories dotted the United States, many of the best in Tampa, Fla., staffed by experienced Cubans. That year sixty cents of every consumer dollar spent on tobacco was spent on cigars.

Thirty years ago, dozens of cigar maker-merchants still lent their fragrance to Manhattan's driven garment district. Annual cigar sales in the United States surged from seven billion to nine billion on the heels of the Surgeon General's 1964 cigarette warning.

The market declined as people realized that cigar smoking was risky too, ebbing at just over two billion in 1993.

Ballyhoo proclaiming the cigar as elite, sexy and necessary to the high life pushed imports to the United States of premium cigars from one hundred million in 1991 to one hundred and thirty million in 1994, to almost two hundred million two years later. That's a lot of fancy cigars, but only a sliver of the total of three billion cigars sold in 1996. How many Americans does it take to smoke three billion cigars? Estimates range from four to eight million.

Yet only a few unpretentious cigar-rolling establishments are left in New York City. Typically, there is room for two or three customers to squeeze in front of the cigar-selling counter. Just a few feet behind the store proprietor or manager, who tends the cash register, sit four or five rollers bent in postures Gompers would recognize. Sometimes a curtain separates the rollers from the storefront. Ventilation, weather permitting, is an open shop door.

The day I visited a storefront back room, workers were constructing Churchills, using Honduran tobacco as filler and binder, Botswana leaf as wrapper. Some scrap—-not much—-went into the rolls. A roller laid two binder leaves on top of each other, scooped up filler leaf, rolled the lot and fitted each "bunch" into a mold. A bunch is an unwrapped cigar. After the stacked molds served short sentences under a bunch press, a specialized staffer removed each bunch individually and wrapped and trimmed it. The cigar was ready to be sold up front.

The bare-handed rollers like

Above: an old-time cigar shop.
Below: hand rolling in New York City.

what they make. Puffing a cigar as one quickly rolls is not easy, but some manage it. Speed is essential since laborers are paid by the piece. The veteran rollers in this shop average two hundred cigars each day, earning about $5 per hour. Rollers are still immigrants: in New York one finds Nicaraguans, Dominicans and Puerto Ricans; in Florida more Cubans. Entertainment to roll by is the radio, turned to Latino music.

A companion pronounced a $1 cigar created on the premises quite acceptable; I found it harsh. But then again, my palate had only lately been educated, or my psyche spoiled, by Professor of Cigars Philip Darrow.

Prof. Darrow teaches Cigar Discernment at one of Manhattan's premier unaccredited institutes of higher education, Club Macanudo. This is a college dedicated to the promotion of the eponymous cigar, manufactured by United Cigar. The club also offers other hand-crafted cigars, major-league drinks, and bar food, served up in the requisite, handsome environment of elaborate ceiling moldings, humidor-lined walls and deep leather chairs—-although its ventilation system isn't quite state of the art. Club Macanudo has prospered sufficiently to spin off a sister club in Chicago and set its sights on Tokyo.

My male and female classmates in

New York included hotel and restaurant professionals and other suits who figured they had to talk the cigar talk, and walk the cigar walk, to succeed. Tuition for the four-class course was $240, which I happily paid. Cigar School has since been collapsed into a less-expensive single session.

"Cigars aren't a health food," Prof. Darrow countered charmingly, when asked if smoking them was risky. Then he understandably changed the subject.

Despite warnings on cigar boxes sold in England and the United States, most cigar smokers underestimate the statistical chance of serious injury while most cigarette smokers overestimate it. Cigar lovers are considerably less prone to lung cancer than those whose pleasure is cigarettes, but still are three times more likely than a nonsmoker to court a malignant tumor where they breathe. The risk of mouth and throat cancers is the same for both smoking groups.

One frequently hears that higher-priced cigars are "organic," that they contain no additives, or many fewer additives, than mass-market cigars. The implication is that costlier is healthier. Frankly, the only certainty is that costlier is more expensive. Once upon a time, it could be said that cigar smoking soothed the psyche of the smoker. That was before the Age of Anxiety about choosing the "right" cigar.

The A.B.C. of a cigar is that it is composed of three elements: filler, binder and wrapper. A cigar in which all elements come from the same place is known as a *puro*.

Puro is not a quality guarantee. Very few cigar-growing areas produce top filler, binder and wrapper. Vuelta Abaio, west of Havana, is one of the rare regions—-and the only one in Cuba—-that grows top-grade tobacco for all parts of a cigar. Tabacalera Fuente created cigar-world news recently when it produced the first Dominican *puro*, Opus X. (The esteemed cigar firm's previous claim to fame was the longevity of its founder, Arturo Fuente, who lived to eighty-five, having smoked twenty-five cigars daily since he was a boy.)

Premium cigars are handmade and all three layers consist of whole leaf, which you can discover for yourself if you care to dissect one. In a nasty cigar, the innards are bits and pieces of leaf. Deconstructed, the tobacco flakes off. The sawdust factor is also attributable to dryness.

The next thing to know about a cigar is that you *can* judge it by its cover. Fifty to sixty percent of cigar flavor derives from its wrapper. What wrapper tastes best is, of course, subjective.

But color is a clue. Cigars with the darkest wrappers are the sweetest smokes. The longer a leaf is in the sun the more time it has to synthesize sugar. *Oscuro*, seldom seen outside Latin America, is the deepest brown, next darkest is *Maduro*.

Dark cigars are sometimes said to taste chocolaty by those who assert expertise. But tobacco doesn't know from cocoa unless someone has been demonic enough to add chocolate specks. If you want chocolate, buy a box of Perugina or a Hershey bar.

Some cigars are dipped in alien flavors, vanilla, for instance, to entice newcomers. I promise you that the candy wears off after a few puffs. You have to like tobacco to appreciate any cigar.

Back to color: caramel-colored wrapper, often with a reddish tint, is known as *Colorado*. Lighter brown leaf with a yellowish hue is known as *Claro*. Brits favor *Colorado*, which is sometimes termed the English Market Selection (EMS), while *Maduro* is the Spanish Market Selection (SMS). Americans prefer the tan *Claro*, dubbed the AMS. *Claro* is pale because it grew up in the shade while *Maduro* was out playing in the sun.

Color terms are often combined. A *Colorado Madura* would have a hint of rust in its dark brown; a

Cigar Manners

- Sniffing a cigar is primitive, and tells you little.
- Squeezing a cigar near your ear could alert you to dryness, but the person who has proffered it to you may be insulted. You can also damage the wrapper.
- Licking a cigar is worse than gauche.
- Teeth do not qualify as a cigar cutter in polite company.
- Retaining or removing the band is up to you.
- Gently brush ash off your burning cigar into an ashtray. Tapping doesn't work well and exposes your cigarette heart. Letting ash drop to the floor reveals you are a slob.
- Do not blow smoke in another person's face, unless the person is an ancient Brazilian warrior who has asked that you puff some courage his way.

Colorado Madura would have a hint of rust in its dark brown; a *Colorado Claro* would be a medium brown tone displaying a trace of red. *Claro Claro* is a bit green.

Colors have sneaky associations. Think of olive oil. Some consumers are drawn to a label marked "light," which they equate with freshness or low fat. But what "light" means is that the oil was obtained after several pressings. Virgin, darker oil, pressed once, is more flavorful. A light cigar wrapper doesn't signify mildness. *Claro Claro* leaves derive their tone from being rapidly cured with heated air. I find *Claro Claro* bitter.

Cigars have two basic dimensions: thickness or ring gauge (a measurement divided into sixty-fourths of an inch) and length. True cigars may be as short as four and a half inches, as long as ten. Standard ring gauges range from thirty-four (Panatella) to fifty-two (Diademus).

Tiny cigars are sometimes called *cigarillos*. Many Europeans smoke cigars so small that

Detail of an ad for Rolando cigars.

as California tobacconist Lew Rothman quipped in a *Smoke* essay, it's a wonder they don't swallow them.

Leave your fashion sense home when you shop for a cigar. A small, slim cigar may appear dainty but probably is stronger than a thick one. More tobacco means more melding and mellowing of flavors. A tough, mean looking cigar may prove a pussycat.

What the length of a cigar tells you is how long it will take to smoke. Don't go for a seven-inch Churchill unless you have an hour to kill or, like Sir Winnie, are prepared to smoke through war and peace.

A large cigar, it is said, is less likely to blister your lips than a small one. Torpedoes and other tapered cigars are prone to hot ends. A century and a half ago, Germans invented a tube, a meerschaum cigar holder, to overcome the difficulty. These days, smokers are willing to leave longer butts or take the heat.

Selecting a cigar shape is opportunity enough to express your personality. Feeling aggressive? Try a pointed Belicoso. Want something unusual? Go for the twisted trio Culebra. Nationality also counts. There is the country of tobacco origin to consider as well as site of manufacture.

The label on a cigar box tells you where a cigar was made, the brand name (which is the name of the producer) and the model. The model designation may be just a number, or the cigar type (Corona), or both, or a flight of fancy.

It sounds straightforward but it's chaos.

Take brands. Ford is Ford in Detroit, Santo Domingo or Paris. Not so with cigars. Partagas, for example, is both the name of a Cuban cigar producer and an unconnected cigar maker in the Dominican Republic. Cuban Partagas are made by a producer designated by the state monopoly. Dominican Partagas are made by a producer working for the American company, United Cigar, which owns the right to the Partagas brand in the United States. Cuban Paratagas cigars and Dominican Partagas cigars are different blends.

Consider now, models: Ford has a successful model named Taurus. Chevrolet may copy the car, but the new Chevy model will be named something else. Not so with cigars. If one producer has a winning model called Taurus, that's the model name a dozen other brands will use. What's more, if Taurus is so popular, why not go for variations on a theme and create Taurus No. 1, Taurus No. 2 and so on? In Cigarland, dozens of dif-

Who chose Sigmund Freud's cigars—his id or his ego?

Growing Blues

Raising and curing cigar leaf is not unlike growing cigarette tobacco, the old fashioned way. But it's more fastidious, takes longer and costs more. A prime tobacco plant yields only sixteen leaves good enough for cigar. Wrapper tobacco is the most expensive crop; the cost can run to $25,000 per acre.

Tobacco for classic cigars is hand-picked green and twined with palm strips, writes Richard Carleton Hacker in "The Ultimate Cigar Book," to which I owe much of this description. The tobacco yellows as it dries on poles in curing barns.

The best cigar tobacco is naturally air dried. From barns, the tobacco moves to packing houses where leaves are sorted by size and texture and tied into "hands." The hands are stacked closely so that little air can circulate, thus promoting the fermentation that will brown them. The leaves are then separated for aging, which may proceed for some months. The tobacco is misted to rehumidify it before stemming. Inspectors grade the leaves as wrapper, filler or binder.

Short of burnng barns, the biggest threats to tobacco lurk in the field. More feared than insect enemies is Blue Mold. If it rains too much in the balmy Caribbean, the combination of heat and excessive humidity can encourage this killer fungus. Much of the Dominican crop of 1985 and the Cuban crop of 1993 fell to Blue Mold. For Cuba, the disaster was Part II of a double whammy. In 1992, Hurricane Andrew wiped out many of the country's tobacco sheds.

Premium cigars on the market reflect the growing conditions of years past. Factories' failure to predict growing label demand is not the only reason you can't always get what you want. Nature may have vetoed your desire before you formed it.

ferent smokes are called Rothschild, Lonsdale, Robusto, etc.

When expert panels rate cigars, they usually do it by cigar type, based on shape and size. So pyramids will be compared to other pyramids. One brand may call its model Pyramid, another may name its pyramid Torpedo. Not all models baptized Torpedo are identically tapered or of an exact size, and tobacco blends may be radically different.

The moral is, that if you find a cigar you like, or wish to give someone a gift of his or her favorites, recalling one key word won't help. You need to memorize or write down: country, brand, model. Got that? Let's move along into a related muddle.

Are Cuban cigars really the best? Aficionados differ in their opinions. Many give an unqualified yes to the celebrated Cohibas and Montecristos. But some think that top Dominican and Honduran cigars are the peers of other good Cubans. The higher value generally put on *Habanos* by Americans, they submit, is connected as much to their relative scarcity— or at least, illicitness—as it is to their quality.

Will the real Cuban cigars please stand up? Some Cuban cigar producers fled to Central America after Fidel Castro's forces confiscated their fields and factories. They brought with them their skills and seeds but they couldn't carry the topsoil.

Is the earth in some quarters of Central America just as good as Cuba's? And what of the adaptation of the plant? Are the differently blended Dominican-made Montecristos the equals of Cuban originals? Are there originals? Every new cigar reflects the state of plant stock and weather conditions of a few years back. Every old cigar also reflects its age.

Argue on—-without me.

Even where it is legal to buy Cuban cigars, famous box markings and band insignias may be forged. Rain spots are sometimes fabricated to make wrappers look natural. The "Cuban" cigar you bought could be a foreign ringer. Or it might be a third-rate Cuban. Not all Havana cigars are created equal—-genuine, trashy Cuban cigars are available on and off the fabled island. When the U.S. trade embargo is canceled, as it will sooner or later be, there will be winners and losers.

Particular cigars from Ecuador, Nicaragua, Germany and the Canary Islands are also touted. Certain manufacturers are highly praised. Some swear by Davidoff, a Swiss company that supervises the making of cigars in the

Dominican Republic and oversees their aging in the Netherlands and Connecticut before they are shipped to be sold. Others swoon over Cameroon wrappers married to Dominican filler in certain Partagas cigars. Some die for Punch.

Upscale smokes made in two dozen nations—-containing tobacco from a growing list of countries—-clutter the international marketplace. The thicket of labels and models vying for attention is terrifying. Even scarier are the nouveau critics, advising that such-and-such a cigar is "creamy." Will the tobacco coat your tongue? Or "leathery." Does it taste like an old shoe? Or "lusty." Is this cigar bent on rape?

In a candid moment, Edgar Cullman Jr., CEO of Culbro, the parent of United Cigar, told *Cigar Aficionado*, "the number of unique tastes that are out there today, are not that significant."

So is all the publicity hype? Ads are beyond suspicion; they're already convicted. But what of the magazine features, the comparative "smokers," the cigar reviews? Yes, much of this is good-for-business hype. But the cacophony of voices is also part of a struggle to craft a helpful vocabulary of taste. Meanwhile, any competent salesperson, given one or two clues about your tastes, can make reasonable suggestions.

My favorite dab of hype—-call it lore if you prefer—concerns ash. The assertions are that whitish, fine ash reveals that the tobacco is well-aged and that the cigar is of superior construction. Reading ash as one would the entrails of a chicken is impractical. Show me the cigar merchant who will allow you to light up a cigar before you show him the money.

A double-guillotine.

Once you've selected your luxury smoke, clipping it is the next challenge. The cut comes at the closed or capped head, the end you slide into your mouth. Most cigars wear their bands as neckties, a couple of inches below the cap. The idea is to slice off the cap, taking as little as possible of the rest of the cigar with it.

Most sharp blades will do. But stay away from any gizmo that only works well with one size cigar. The double guillotine is stylish but a cigar scissors is easier to manipulate. I'm suspicious of double-use instruments—-the lighter with the hidden blade, the piercing pen. They reek of Swiss army-knife (SAK) syndrome. If you're stranded in the jungle with a corked wine bottle, a can of beans and canvas you need to convert to bandage, a SAK is handy. But it does no job well.

Sterling implements may dull in time, so why splurge? It's better to select a cutting edge that it's okay to lose.

Allow the seller to clip the cigar for you only if you intend to smoke it right away. Cigars dry out very quickly.

Choose your smoking venue carefully. The cigar police are everywhere.

The foot of a cigar, the end which is lit, is called the tuck. Hold the cigar at a forty-five degree angle from a long, lit match or a butane lighter. Other lighters may emit fumes which will taint the tobacco. Rotate the tuck in the flame until it is evenly lit. Puff. Patience is required. Among the teachings of Prof. Darrow is that it often takes three matches to light a cigar.

Puff your lit cigar, hold the smoke in your mouth a bit if you like living dangerously, but don't inhale it. You can grasp a cigar any way you like. But I must warn you that cigar snobs disdain those who handle it like a cigarette. If your pleasure is interrupted, set the cigar in an ashtray but don't stamp it out. If the fire expires, no penalty is attached to relighting it.

Cigars to be smoked in the future are happy in a dark, moist cocoon. Pig bladders once filled that role. But the animal-rights crowd and various health squads would be down on that sort of thing these days, so if you get serious it's advisable to devise or purchase a humidor.

The mantra of successful storage is seventy-seventy, Fahrenheit degrees and humidity reading. Ballpark figures are close enough. Room temperature is the right temperature in which to keep a humidor, unless you live in an igloo.

Cellophane wrap, cardboard boxes or metal tubes won't do the trick. The problem with such humble storage is not the humility, it's the humidity. In fact, an adequate humidor can be fashioned from Tupperware and a sponge of florist's oasis soaked in distilled water.

Tap and spring waters are taboo because they contain metallic traces that will seep into the cigars, just as open tins spoil the taste of food.

A fine humidor need not be made of cedar, but thin cedar dividers will prevent the flavor of one cigar from corrupting its neighbor. A functioning hygrometer is necessary; a simple hardware-store appliance is as good as any other.

The crass humidor question is: how high is up? The richest price ever paid was for John F. Kennedy's, auctioned in 1996. You can see a picture of it in the chapter, "Tobacco Treasures."

A Cuban tobacco chest, signed by Fidel Castro and stocked with one hundred and fifty commemorative Partagas, also brought big bucks. The buyer was the French government tobacco monopoly, Seita, which anted up $67,000.

Still for sale is the limited-edition mahogany "Millennium," offered by the Swiss firm, Perrenoud. Sales literature promises that this computer-controlled storage unit for fifteen hundred cigars was executed by remarkable craftsmen from a design by engineers and physicists. How many are rocket engineers is not noted. The oval humidor is leather-trimmed and its hinges gold-dipped. At $150,000, it's not your father's whiskey locker.

If the Millennium is sold out by the time you read this, there are hundreds of other humidor choices, including those in the three-figure category.

Cigars continue to mature—which means decay—in a humidor, a plus for most varieties. *Smoke* writer Adrian Bartoli compares the process to the making of an Egyptian mummy. There's actually quite a chemical brawl going on in the cigar closet as each cigar expels ammonia, and reactions rage among hydrocarbons, sulfur and oils. Bacteria are at work. Cigars rot for your pleasure.

Two or three years after being made, the full flavor of most cigars has been developed. After that, mustiness may set in, still some folks lap up musty.

Don't stick your sticks away and forget about them. The interior conditions of most humidors need regular monitoring. One of the really bad things that can happen to stored cigars is an infestation of lasioderma. The insidious tobacco beetle lays its microscopic eggs on the leaves of growing plants.

TOBACCO ENEMY

The hairy lasioderma grub chomps on stored cigars.

When nobody's looking, the eggs can hatch into larval grubs who contentedly munch through your Cohibas on their way to beetlehood.

Nineteenth century newspaper publisher Horace Greeley infamously described a cigar as "a fire at one end and a fool at the other."

I wouldn't go that far. But with cigars, as with other things worth having in life, persnickety preferences can lead to a lot of trouble. If you like the idea of a cachet cigar more than you like actually smoking one, you may not want to strive for superior taste. At the very least, consider whether there is time enough in your life for a second career.

If, however, you love the firsthand smoke of a good cigar, your work is cut out. Good luck.

'Visitors are implored to squirt the essence of their plugs in the spittoon!'

Charles Dickens
American Notes, 1842

Chew This Over

Chew may be lovelier when package stays sealed.

IF ive million Americans chew tobacco. You don't want to be in spitting distance of any of them.

Actually, I'm lying—not about the distance, but about the number. The stats don't distinguish between chewers and dippers. Go back to "The Right Snuff" and you'll see that dippers suck moistened powdered tobacco, a variety of snuff. But chewers want something they can get their teeth into. If you're interested in chew, a.k.a chaw, a.k.a. eatin' tobacco, you're in the right place.

Blame it on the Navy. From Columbus' day on, sailors lobbed gobs of stringy tobacco into their mouths for the same reason some Native Americans did: to fight hunger or fatigue. If all you have on board are some weevily biscuits, could be that tobacco is not going to taste that bad. If you're nodding off at the rudder, tobacco is going to snap you awake.

Chewing spares you the inconvenience of a pipe, or fooling with fire in a thirty-knot wind. You can chew and do something energetic at the same time, such as swabbing the deck, although chewing (which can lead to dribbling) is going to make your work harder. Can you imagine someone hurling a hundred-mile-per-hour fast ball with a stogie in his mouth, or sliding into home plate, inhaling a Camel?

Chewing is no more unsightly than the mumps. And if you're in the middle of the Atlantic, or in the Yankee outfield, chances are you can squirt off your juice without it darkening someone else's eye.

Chewing used to be more up-close and personal than it is now. Seventeenth-century gents who indulged carted around silver spitting pots. That's how the spittoon got its start.

Before televised football came along, chewing was a prime-time American sport. Men's clubs and other male precincts had spittoons, a.k.a. cuspidors, as permanent installations. But not enough of them, according to the otherwise intrepid traveler, Charles Dickens.

Dickens knew the filthy slums and workpits of England. He had observed the lowest of the low. He knew dirt. But, as he recounted in his 1842 *American Notes*, he'd never seen anything resembling the degradation of Washington, D.C.

"Washington may be called the headquarters of tobacco-tinctured saliva," his travelogue reported. "The prevalence of those two odious practices of chewing and expectorating...became most offensive and sickening."

Chewing was not confined to the political haunts of the Potomac. Dickens continued, "In all public places of America this filthy custom is recognized. In the courts of law the judge has his spittoon, the crier his, the witness his, while the jurymen and spectators are provided for, as so many men who in the course of nature must desire to spit incessantly.

"In the hospitals the students of medicine are requested by notices on the wall to eject their tobacco juice into the boxes provided for that purpose and not to discolor the stairs.

"In public buildings visitors are implored to squirt the essence of their quids, or 'plugs,' as I have heard them called by gentlemen learned in this kind of sweetmeat, in the spittoon, and not about the bases of the marble columns."

Dickens came across the worst on a ferry crossing enroute to Southern plantations. He found no position on the deck safe from the tobacco spittle which besmirched his fur coat.

He also visited a tobacco factory in Richmond, Va. where he "saw the whole process of picking, rolling, pressing, drying, packing in casks and branding...One would have supposed

Vintage ad from days chew cleverly announced itself.

there was enough in that one storehouse to have filled even the comprehensive jaws of America."

Not by a long shot. A lucky mutation of Ohio Valley tobacco into a sweet White Burley in the 1860s translated into a chaw American males could not resist. In 1885, the small city of Danville, Va., housed twenty-six

THE SMOKING LIFE 145

tobacco manufacturers. Twenty-four of them were devoted to chew. In chewing's peak year, 1890, thirty-one pounds per capita of chaw were produced in the United States.

Varieties of chew then included Flat Plug (compressed rectangular cakes of Bright tobacco), Navy (compressed Burley), Twist (tobacco twined into a rope and then compressed) and Fine Cut (shredded but not pressed).

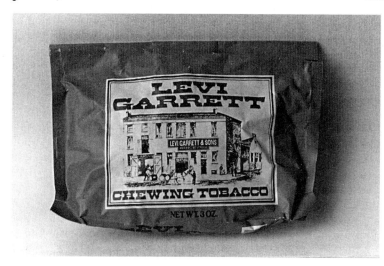

Contemporay chew package harks back to glorious past.

Today's labels are less informative. In the interest of art, if not science, I chomped a wad of the contents of a bag marked "chewing tobacco." It was like biting off the end of a damp mop. No amount of sweetener could make me forgive its texture or accept its bitterness. There's a buzz, but so what?

In my view, chaw, like nicotine gum, is a substance more suited for aversion therapy than pleasure. Chewing is an experience I'm likely to repeat only if invited to pitch the first baseball at the World Series.

I'd advise anyone about to undertake a similar experiment to have a disposable container at the ready. Fewer public spittoons are available to receive chewers' spent pleasures than existed in Dickensian America.

In 1955, the American Tobacco Company had a chew-related problem. According to a company memo, since it had become impossible to find individuals willing to clean the cuspidors in its New York office, the decision was taken to remove them. In 1997, when I visited an R.J.

Reynolds cigarette factory in Winston-Salem, N.C., there were spotless ashtrays at every turn, but nary a spittoon in sight.

Chewing as a pastime has its risks, even for strong jaws. As long ago as 1795, English physician Benjamin Rush spotted a correlation between tobacco chewing and injury to the mouth and stomach. Current epidemiology tells us that three to five per cent of smokeless-tobacco users will be stricken with mouth cancer; more will be plagued by periodontal disease.

The art of chewing is thought of today as a redneck pleasure, although there are reports of college boys taking it up in lieu of smoking. On close examination, most frat-house chew turns out to be mealy-mouth dip. Eating tobacco apparently requires greater resolve than most university types can summon.

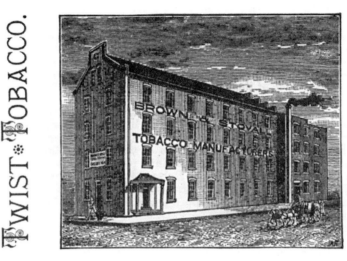

Danville, Virginia lived on chew; a factory circa 1885.

When I think of the hardships committed chewers face, and look back on my own little chaw adventure, my distanced esteem for the breed only increases.

'May you smoke? What's *that* supposed to be — homage to a lady?'

Grace Kelly
The Country Girl, 1954

Lighting Up the Screen

The fantasy world of our time is the cinema. Television is not even a close second: the box is crammed with too much that is depressingly real and too much that is patently false. Even the best of its images are smaller than life. Films, however, are magical, and their magic is permeated with smoke. To smoke is both to escape and to be present more fully. It's a contradiction every smoker knows. That first puff or two is a fleeting diversion from the here and now, giving us precious seconds to collect our thoughts, a prelude to sharpening our focus. Watching a movie is a comparable sensation. We sit back, relax and then, if the magic is working, we are acutely in the world of the big screen.

That world is smoky. Once upon a time when the projector's light cut through the theater air, motion pictures announced themselves in a smoke-like haze. And it was a sad movie palace that did not boast a loge, where patrons could join the characters on

Rita Hayworth in "Gilda," Columbia, 1946.

screen in a smoke. The cigarette, especially, is an accouterment of elegance as well as a prop of grit. Only rarely does smoking call attention to itself in old movies; rather it is an almost omnipresent element of romance and struggle alike. We more frequently see smoking than eating, dressing or even undressing. Smoking is the emblem of social interaction, polished or grimy, and of the individual alone in thought.

All those nicotine-loving movie stars didn't make me wanna do it. I was a smoker

The "Now Voyager" love affair ignites. It will smolder when Henreid asks, "Shall we just have a cigarette on it?" Warner Brothers, 1950.

before I was a moviegoer, and most of my favorite smoking scenes are in films made before I was born. Smoking is integral to each of them; none could be successfully re-imagined without the smoke.

Winner in the Romance Category is the glittering night in *Now Voyager* when Jerry, played by Paul Henreid, places two cigarettes between his lips, one for himself and one for the woman (Charlotte) he can't keep, played by Bette Davis. Refusing to make of impermanence a tragedy, Davis clinches the film with "Jerry, don't let's ask

150

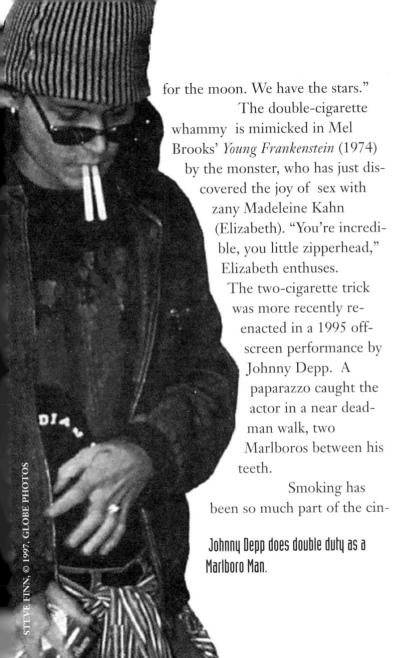

Johnny Depp does double duty as a Marlboro Man.

for the moon. We have the stars." The double-cigarette whammy is mimicked in Mel Brooks' *Young Frankenstein* (1974) by the monster, who has just discovered the joy of sex with zany Madeleine Kahn (Elizabeth). "You're incredible, you little zipperhead," Elizabeth enthuses.

The two-cigarette trick was more recently re-enacted in a 1995 off-screen performance by Johnny Depp. A paparazzo caught the actor in a near dead-man walk, two Marlboros between his teeth.

Smoking has been so much part of the cinematic world that characters managed it (or hoped to) even under extraordinary conditions. Lon Chaney, starring as an armless strongman in the classic silent, *The Unknown*, adeptly smoked cigarettes he held between his toes. William Bendix, having suffered a leg amputation while stranded on Alfred Hitchcock's *Lifeboat*, announces he'd trade another limb for a smoke. Having just captured *Iwo Jima*, John Wayne (a cigar enthusiast in real life) declares, "I never felt so good in my life. How about a cigarette?" He is taken out by a sniper's bullet as he lights up.

The envelope please. Winner for Funniest Smoking Picture is *The Mask* (New Line, 1994). In his dreams, Jim Carrey (Stanley) lights a cigarette to land Cameron Diaz (Tina). In his mask, Stanley corners Tina and demands, "Want a cigarette? No?" He blows a smoke heart and shoots an arrow at her through it.

Stanley's buddy, played by Richard Jeni, is on the prowl, too. He stops a cigarette girl in a smoky casino: "Give me a pack of cigarettes."

"Which brand, sir?"

"It doesn't matter. I don't smoke. But for you I would shoot the Surgeon General."

At the film's climax, Tina is tied to a bomb, seconds away from exploding when the masked Stanley comes to the rescue. In his moment of victory, he switches to a cigar.

A couple of years ago, a pair of Paul Auster/Wayne Wang features, *Smoke* and *Blue in the Face*, starred smoking as a way of life. Both are anchored by Harvey Keitel (Augie), as manager of a scruffy Brooklyn tobacco shop. In *Smoke*, customer William Hurt tells Augie how Sir Walter Raleigh weighed smoke. In the sequel, the hole-in-the-wall smokers' hang-out appears to be threatened. "You mean the Brooklyn Cigar Company's gonna become a health food store?" Augie bellows at its owner.

But we sense the Brooklyn Cigar Company will scrap on, dispensing cigarettes, cheap cigars and sweet memories. "Sharin' a cigarette with your lover..." reminisces Jim Jarmusch.

"That's bliss," Keitel smiles.

Their cigarettes remain as true to them as Bogie's did after Ingrid Bergman boarded that plane leaving Casablanca.

Most Hollywood movies, if not European ones, are less smoky than they once were. Yet the cinematic cigarette is not yet consigned to marginalia. Look and you'll see the cigarette is still there—-identifying rebels and marking wayward women, playing the same roles in life and cinema it did when the Silents yielded to Talkies. And cigars remain icons of power. On screen or off, Hollywood has always loved tycoons.

No catalogue could be compiled of every larger-than-life tobacco moment in film. But here are some of my favorites, capturing movieland at its smoking best.

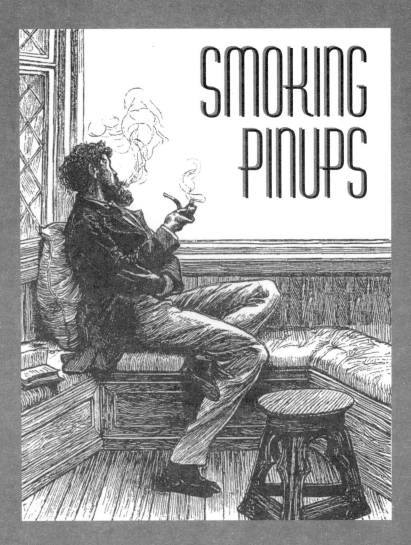

BEST SMOKING PICTURE... 'ROBERTA'

When a lovely flame dies/Smoke gets in your eyes.
—Jerome Kern and Otto Harbach, 1933

Roberta, a Broadway musical before it went to screen, gave us the song with the greatest smoke lyric of all. The theatrical director wanted to cut *Smoke Gets in Your Eyes*, but Kern resisted. Irene Dunne sings it as cigarette smoke swirls round her. But Dunne doesn't smoke; it's a metaphor. Swinging into the breach are Ginger Rogers—on the cigarette, and Fred Astaire—on the pipe. Naturally, there's some fancy footwork, too. But not before Rogers protests, "I won't dance/Don't ask me."

The last musical bonbon of this vintage RKO film is *Lovely to Look At*. And so the movie is. The action is centered in a Parisian couture house; the glamorous clothes and the sets are art deco to the nines. Even if *Roberta* weren't perfumed in smoke, I'd love it.

Marlene Dietrich: "There is a foreign legion of women, too. But we have no uniform, no flag and no medals." *Morocco*, 1930. Paramount

Carole Lombard (Lily Garland): "You forget who I am."
Twentieth Century, 1934 Columbia

Jean Harlow smokes and James Cagney steams.
The Public Enemy, 1931. Warner Bros.

Jean Gabin: "English tobacco hurts my throat. It seems your gloves, your tobacco, everything comes between us."

La Grande Illusion, 1937. Réalisations Art Cinématographique

Tallulah Bankhead puffs it up in *The Devil and the Deep*, 1932. Paramount

Claudette Colbert smoked with style.

William Bendix: "Right now I'd give the other leg for a cigarette." *Lifeboat*, 1944. Twentieth Century Fox

Lauren Bacall (Slim) turned "Anybody got a match?" into the sexiest entry line in film history.

To Have and Have Not, 1944. Warner Bros.

Myrna Loy, Melvin Douglas and Cary Grant (Mr. Blandings): "I've been looking forward to sitting in front of the fire with my pipe and my new smoking jacket."

Mr. Blandings Builds His Dream House, 1948. RKO

Jane Wyman: "I've only played one role. It was in a church hall. I played the fourth deadly sin."

Stage Fright, 1950. Warner Bros.

The once and future President in his movie days.

The scene: A hospital where Marlon Brando (Ken), a WWII paraplegic, has wrenched himself up, then collapsed.
Jack Webb (Norm): "Where did you think you were going?"
Brando: "I want a cigar store."
Webb: "You better start with cigarettes."

The Men, 1950. Republic Pictures

Leslie Caron (Gigi):
"Everything is love...An ugly, black cigar is love."
Gigi, 1958. MGM

The lady poet: "Why think about the future...I know exactly what I like best, the three escapes: smoking, drinking and bed."

La Dolce Vita, 1960. Rima/Pathe

Marilyn Monroe (to Tom Ewell): "With a married man it's all so simple. It can't possibly get drastic."

The Seven Year Itch, 1955. Twentieth Century Fox

Audrey Hepburn (Holly Golightly): "I honestly think I'd give up smoking if he asked me."

Breakfast at Tiffany's, 1961. Paramount

Sharon Stone (Catherine): "What are you going to do? Charge me with smoking?"

Basic Instinct, 1991. Guild/Carolco

Jack Nicholson: "You went to a doctor — why?"
Streisand: "Because five packs a day isn't normal."

On a Clear Day You Can See Forever, 1970. Paramount

"Tobacco inspires courage of another kind, deliberate yet immovable, affectionate and feeling, yet despising danger.

Daniel Webster

Smoking Guns

An army runs on tobacco. This was true for Native Americans who had a war pipe as well as a peace pipe.

The Spanish, bent on conquering Mexico, were quick to see the advantage smoking gave the Aztecs and copied the tactic. Juan de Cardenas, who accompanied Cortez, observed, "Soldiers subject to privations, keep off cold, hunger and thirst by smoking."

Europeans adopted this wisdom on home soil during the Thirty Years War, which began in 1618 and eventually involved France and much of Scandinavia along with German and Middle European states. Their smoking gun against hardship was the pipe.

Louis XIV seized upon the lesson and ordered a regular issue to French soldiers of a pipe, tobacco and, for lighting up during the winds of war, flint and steel.

Tobacco was also key to troop R & R. This sergeant's mess song is centuries old:

> So we'll smoke our pipe
> And we'll drink our glass,
> And we'll play our game
> And we'll hug our lass.
> And as for the rest—
> Why the devil's an ass!
> Drink boys, drink!

As the drum beat for American independence, Gen. George Washington beseeched the Continental Congress to send tobacco for his revolutionary troops. Ben Franklin in Paris, looking for funds to finance the rebellion, borrowed two million livres against a promise to deliver five thousand hogsheads (barrels) of Virginia leaf. Thus was tobacco an underwriter of American liberty.

The Redcoats smoked, too, and did not lose their taste for tobacco when they lost the war. Generations of smokers later, Lord Wellington tried to pacify Queen Victoria by barring pipes and cigars from the mess halls of the lower ranks. He dared not alienate his officers by requiring a similar sacrifice.

In the U.S. Civil War, tobacco was part of the food ration for Southern troops. Robert Mayo, a Richmond tobacco merchant, kept the Confederate navy in plug, while James A. Thomas, an advisor to Jefferson Davis, supplied the army, often at his own expense.

The tobacco tax, instituted by the Union to pay for its cause, rocketed up prices on manufactured smokes. The levy was eight-tenths of a cent on a penny cheroot, four cents on a nickel cigar and $2 on one thousand cigarettes. Fans kept Ulysses S. Grant in cigars, but his foot soldiers fended for themselves. The aftermath of more than one bloody battle was a trade fair, Union Blues bartering coffee for the Southerners' tobacco.

On April 2, 1865, a week before Appomatox, Confederate troops fleeing Richmond set its tobacco warehouses afire to keep the good stuff out of enemy hands. Their spite had the unintended consequence of incinerating most of what had been the Confederate capital.

When the smoke had cleared, tobacco had both won and lost. Much of Tobaccoland was in embers, fields and storehouses included. And American tobacco taxes were here to stay. Tobacco lust had soared, and a huge industry, led by rags-to-riches North Carolinians Buck Duke

> **Cradle of Demos-smoke**
>
> *Greece, where democracy began, has more smokers per capita than any other European nation. In 1995, Greek authorities launched an expensive anti-smoking campaign, only to admit a year later it had utterly failed. More Greeks smoked after the campaign than before.*

Ils fument!

and Robert Joshua (R.J.) Reynolds, rose from the ashes.

In Europe, war would also bring about change, tobacco-wise. As English and French officers had been won over by the cigar in dust-ups with Spain, military adventurism in the east would make the cigarette the winner of the Crimean War. Officers doted on the exoticism of Turkish tobacco and the elegance of cigarette holders.

The victory of the cigarette was clinched by World War I. Combatant governments realized that their armies could not fight without it. In the trenches of World War I, the cigarette shed the vestiges of its effete image and emerged as the gutsy smoke.

Belgian soldiers cabled the Minister of War: "Give us less food if you like, but let us have tobacco."

The British fighting man had a blunt motto: "Smile while you've got a lucifer [wooden match] to light your fag."

General John Pershing, commander of the American Expeditionary forces, told Washington: "You ask me what we need to win this war. I answer, tobacco as much as bullets."

U.S. canteens began offering free cigarettes to service personnel along with donuts and coffee. The YMCA kicked over the puritanical traces when it joined in. The volunteer Army Girl's Transport Tobacco Fund sent millions of cigarettes overseas. Pressure on the Pentagon to supply cigarettes increased as medics in the field joined the chorus demanding regular supply. "As soon as the lads take their first whiff," a surgeon reported from France, "they seem eased and relieved of their agony."

Pershing again deemed tobacco "indispensable." In 1918, the U.S. War Department made tobacco part of the daily ration.

Both Germany and the allies supplied troops with tobacco in various forms. But most soldiers were happiest with cigarettes. Nicotine—-the miracle drug, both stimulant and soother—-won the day.

The cigarette accompanied survivors home. Like satellite communication in our time, what had been a military toy became a civilian joy.

By World War II, most soldiers could be said to be carrying their habit overseas as opposed to bringing it back with them. President Roosevelt declared tobacco an essential wartime crop.

The good guys were tobacco lovers. FDR, Winston Churchill and General Charles de Gaulle brandished their respective smokes at almost every photo op. Adolph Hitler, the queasy little fuhrer, detested both cigarettes and meat.

Smoking reflected the good character of the American soldier. An army training manual by Edward Andrews, printed at the height of the war, named the requirements of leadership:

- Be calm in an emergency.
- Give clear orders.
- Smoke and make your troopers smoke.

Andrews recommended smoking because it killed time but kept the soldier alert.

Gen. Dwight D. Eisenhower led the charge, smoking like a volcano. (Mamie kept the

home fires burning, the better to light the cigarette she almost always had in hand.)

The grunt's K-pack contained Spam, soap, dehydrated lemonade and four Chelsea cigarettes, not nearly enough to fuel the average G.I., who consumed thirty cigarettes each day.

Soldiers cared what they smoked. General Douglas MacArthur barked at the Pentagon for stocking up on off brands or, worse yet, assuming soldiers could get by with local crud. "My soldiers need American tobacco," he declared.

They got it—-as part of their rations, and at cost from the army PX.

U.S. tobacco companies were delighted. The American Tobacco Co. had another coup with Lucky Strike. Before the war the package was green, but the chemical that made it green turned out to be needed for the war effort. So the circle went red, accompanied by the advertising slogan: Lucky Green Has Gone to War.

It got so that Luckies, dressed in any color, and other standbys—-Pall Malls, Old Gold—-were hard to come by on the homefront. Wartime cigarettes included Ramsey and Maypo, made of tobacco so American it had been doused in maple syrup.

Cigarettes clinched the victory, or so it's possible to believe, and kept the postwar peace. For two years after VE day, American cigarettes constituted the most stable currency of

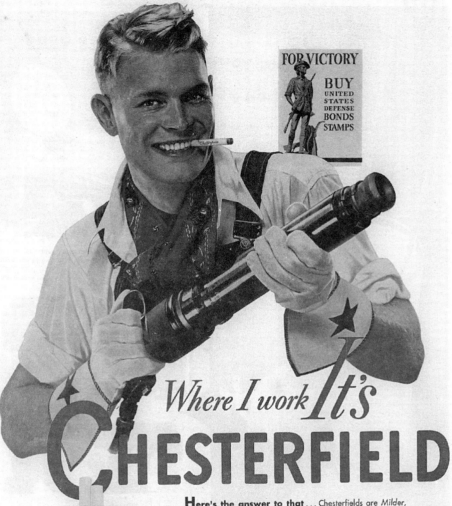

occupied Europe.

    John Tate of Sutherlin, Va., who soldiered in that war, was a rarity. He didn't smoke—-disliked cigarettes, although he enjoyed the occasional cigar or pipe. Odder still, Tate was the son of tobacco growers. Don't ask!

    In the waning days of the war, as his unit shuffled between Germany, Luxembourg and Belgium, Tate picked up his regular fifty-cent cartons from the PX. He then sold them to the Milo Minderbinder of his unit, a wheeler-dealer who frequently made a run to Paris to supply the local—-color it black—-market. Tate recalls getting as much as $100 per carton; the middleman's cut was ten percent. "Cigarettes over there were priceless," Tate said, who after the war ran a general store for fifty years but never again benefited from such a mark-up.

    In postwar Germany, Camels, Chesterfield and Lucky

Strike—had astronomical barter value. Ten cigarettes bought a restaurant meal, or a night with a prostitute, with two more thrown in for a tip. Connoisseurs of lasting objects could score a silver tea service for two cartons; a painting for twenty.

In 1947, Hamburg police offered a reward of five cartons of American cigarettes for information leading to the arrest of a murderer.

U.S. might smoked its way to stalemate in Korea and Vietnam. But the smoking guns of America stopped blazing a couple of years ago. The Department of Defense, forgetting what had once made America strong, banned smoking in several sections of military installations. It followed this up with an order that commissary cigarettes be sold at market price.

Military morale plunged, some say. Maybe "peace keepers" can get by without tobacco. But can warriors?

'With pipe and book at close of day, Oh what is sweeter, mortals say.'

Richard le Gallienne

Literary Puffs

While tobacco-smoke for countless centuries was a lyric in the spoken poetry of Native Americans, it took nearly a hundred years for it to mist its way into European verse. There may have been earlier ditties now forgotten, but on the extant evidence, tobacco made its western literary debut in 1590: In Spenser's *Faerie Queene* the muse, Belphoebe, prepares "divine tobacco."

The phrase is echoed in a 1599 poem by the Earl of Essex who rhapsodizes the "sacred flame" as the antidote to a troubled spirit.

Not a half-dozen years later, *Don Quixote* tilts at a jailer's pipe in Cervantes' Spanish epic. The mention is so off-hand as to insinuate that smoking is so ordinary that nothing more need be said about it.

For most of the last three hundred years, tobacco has casually scented much of western literature. Most notes of defensiveness are either recent or truly antique.

The first excerpts here capture the wonder of tobacco when Europe was old but smoking was new. The later ones I love for the mood each evokes, for the window each opens on its time and characters.

THE HEALING HERB

Pandora (to servant)
(after accidentally spearing her lover)

Gather me balms and cooling violets,
And of our lovely herb Nicotian,

And bring withal pure honey from the hive
To heal the wound of my unhappy hand.
—*The Woman in The Moon*, 1597
by William Lilly

IF IT WALKS LIKE A DUCK

<u>Bobadil</u> (to Hercules)
Your Nicotian is good too, I could say what I know of the virtue of it, for the expulsion of rheums, raw humors, crudities, obstructions and thousands of this kind, but I profess myself no quack.
—*Every Man in His Humor*, 1598
by Ben Jonson

THEORY OF EVOLUTION

Blessed age, wherein the Indian Sun [tobacco] has shined.
Whereby all arts, all tongues have been refined...
The daintiest dish of a delicious feast
By taking, which differs man from a beast.
All gods, all pleasures it in one doth link
Tis Physics, clothing, Music, meat and drink.
—*The Metamorphosis of Tobacco*, 1602
by John Beaumont

AN OLD-FASHIONED FELLOW

While all the world was snuffing my uncle Toby continued to smoke his pipe, thus displaying in his unobtrusive manner his scorn of the daintier Nicotian custom.
—*Tristram Shandy*, 1760
by Laurence Sterne

MANHATTAN MYSTERY

It has been traditionary in our family that when the great navigator [Henry Hudson] was first blessed with a view of this enchanting island, he was observed for the first and only time in his life to exhibit strong symptoms of astonishment and surprise.

    He uttered these remarkable words while he panted toward this paradise of the new world: "See! There!"—-and thereupon, as was always his way when he uncommonly pleased, he did puff out such clouds of dense tobacco smoke that in one minute the vessel was out of sight of land.

—*A History of New York*, 1809
by Washington Irving

EVERY MAN'S PLEASURE

Sublime Tobacco! which from east to west
Cheers the Tar's labor or the Turkman's rest...

Divine in hookahs, glorious in a pipe
When tipp'd with amber, mellow rich and ripe...

Yet thy true lovers more admire by far
Thy naked beauties—-Give me a cigar!

—*The Island*, 1823
by Lord Byron

JANE EYRE'S FLAME

    She [the gypsy fortuneteller] drew out a short, black pipe and lighting it, began to smoke. Having indulged a while in this sedative, she raised her bent body, took the pipe from her lips, and while gazing steadily at the fire, said very deliberately:

    "You are cold; you are sick; and you are silly."

    "Prove it," I rejoined.

    "You are cold because you are alone; no contact strikes the fire from you that is in you. You are sick because the best of feelings, the highest and sweetest, keeps far away from you. You are silly, because suffer as you may, you will not beckon it to approach; nor will you stir one step to meet it where it awaits you."

    She put her short, black pipe to her lips and renewed her smoking with vigor...

    "Rise, Miss Eyre: leave me; the play is played out."

    Where was I? Did I wake or sleep...Again, I looked at the face, which was no longer turned from me—-on the contrary, the bonnet was doffed, the bandage displaced, the head advanced.

"Well, Jane, do you know me?" asked the familiar voice.

"Only take off the red cloak, sir and then—"

"But the string is in a knot—-help me."

"'Break it, sir."

'There, then—-Off ye lendings!" And Mr. Rochester stepped out of his disguise.
—*Jane Eyre*, 1847
by Charlotte Bronte.

BECKY TESTS A CIGAR

Miss Sharp loved the smell of a cigar out of doors beyond everything in the world—- and she just tasted one in the prettiest way possible, and gave a little puff and a little scream and a little giggle, and restored the delicacy to the captain.

—*Vanity Fair*, 1848 by William Thackeray

ROUGH SEAS AHEAD

When Stubb had departed, Ahab stood for a while leaning over the bulwarks; and then, as had been usual with him of late, calling a sailor of the watch, he sent him below for his ivory stool and also his pipe. Lighting the pipe at the binnacle lamp and planting the stool on the weather side of the deck, he sat and smoked.

How could one look at Ahab, seated on that tripod of bones, without bethinking him of the royalty it symbolized? For a Khan of the plank, and a king of the sea and great lord of Leviathans was Ahab.

Some moments passed, during which the thick vapor came from his mouth in quick and constant puffs, which blew back against his face. "How now," he soliloquized at last, withdrawing the tube, "this smoking no longer soothes. Oh, my pipe! Hard must it go with me if thy charm be gone! Here I have been unconsciously roiling, not pleasuring—-aye, and ignorantly smoking to windward all the while; and with such nervous whiffs, as if, like the dying whale, my final jets were the strongest and fullest of trouble.

"What business have I with this pipe? This thing is meant for sereneness, to send up mild white vapors among wild hairs, not among torn iron-grey locks like mine. I'll smoke no more!"

He tossed the still-lighted pipe into the sea. The fire hissed in the waves; the same instant the ship shot by the bubble the sinking pipe made. With slouched hat, Ahab lurchingly paced the planks.
—*Moby Dick*, 1851
by Herman Melville

THE PIPE: INSIDE STORY

I am a writer's pipe. One look at me
and the coffee color of my Kaffir face
will tell you I am not the only slave:
my master is addicted to his vice.

Every so often he is overcome
by some desire or other, whereupon
tobacco clouds pour out of me as if
the stove was kindled and the pot put on.

I wrap his soul in mine and cradle it
within a blue and fluctuating thread
that rises out of my rekindled lips
from the glowing coal that brews a secret spell.
He smokes his pipe, allaying heart and mind,
and for tonight all injuries are healed.
—*Les Fleurs du Mal*, 1857
by Charles Baudelaire

HUCK PIPES UP

The Red-Handed [Huckleberry Finn] made no response, being better employed. He had finished gouging out a cob, and now he fitted a weed stem to it, loaded it with tobacco, and was press-

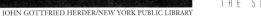
JOHN GOTTFRIED HERDER/NEW YORK PUBLIC LIBRARY

Mark Twain only smoked "one cigar at a time," thirty times each day. In between, he relaxed with a pipe.

drowsiness began to steal upon the eyelids of the little waifs. The pipe dropped from the fingers of the Red-Handed, and he slept the sleep of the conscience free and the weary.
—*The Adventures of Tom Sawyer*, 1876
by Mark Twain

MARRIED TO HIS CIGAR

"You must choose between me and your cigar!"

Open the old cigar box and get me a Cuban stout,
For things are running crossways, and Maggie and I are out.

We quarreled about Havanas—-we fought o'er a good cheroot,
And I know she is exacting and she says I'm a brute.

ing a coal to the charge and blowing a cloud of fragrant smoke—-he was in the full bloom of luxurious contentment. The other pirates envied him this majestic vice and secretly resolved to acquire it shortly.

Presently Huck said: "What does pirates have to do?"

Tom said: "Oh they just have a bully time—-take ships, and burn them and get the money and bury it in awful places in their island where there's ghosts and things to watch it, and kill everybody in the ships—-make 'em walk a plank."

...Gradually their talk died out and

Open the old cigar box—-let me consider a space;
In the soft blue veil of the vapour musing on Maggie's face.

Maggie is pretty to look at—-Maggie's a loving lass,
But the prettiest cheeks must wrinkle, the truest of loves must pass.

There's peace in a Laranaga, there's calm in a Henry Clay,
But the best cigar in an hour is finished and thrown away—-

Thrown away for another as perfect and ripe and brown—-
But I could not throw away Maggie for fear o' the talk o' the town...

Open the old cigar box—-let me consider anew—-
Old friends, and who is Maggie that I should abandon you?

A million surplus Maggies are willing to bear the yoke!
And a woman is only a woman, but a good cigar is a Smoke!
—*Departmental Ditties*, 1890
by Rudyard Kipling

A SUITABLE PROFESSION

<u>Lady Bracknell</u>
Do you smoke?

<u>Jack</u>
Well, yes, I must admit I smoke.

<u>Lady Bracknell</u>
I'm glad to hear it. A man should have an occupation of some kind.
—*The Importance of Being Ernest*, 1895
by Oscar Wilde

SMOKE AND SEDUCTION

Madame Olenska did not move when he came up behind her, and for a second their eyes met in the mirror; then she turned and threw herself into her sofa corner, and sighed out: "There's time for a cigarette."

He handed her the box and lit a spill for her; and as the flame flashed up into her face she glanced at him with laughing eyes and said: "What do you think of me in a temper?"

Archer paused a moment; then he answered with sudden resolution: "It makes me understand what my aunt has been saying about you."

"I knew she had been talking about me. Well?"

"She said you were used to all kinds of things—-splendors and amusements and excitements—-that we could never hope to give you here."

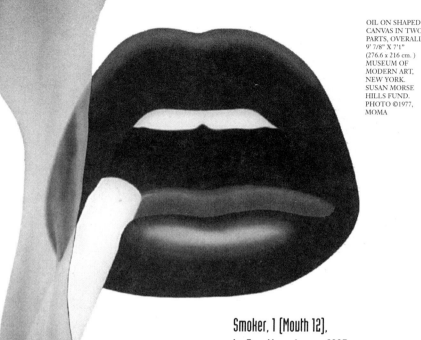

OIL ON SHAPED CANVAS IN TWO PARTS, OVERALL 9' 7/8" X 7'1" (276.6 x 216 cm.) MUSEUM OF MODERN ART, NEW YORK. SUSAN MORSE HILLS FUND. PHOTO ©1977, MOMA

Smoker, 1 (Mouth 12), by Tom Wesselmann, 1967

Madame Olenska smiled faintly into the circle of smoke about her lips.
—*The Age of Innocence*, 1920
by Edith Wharton

THE MAGIC CIGAR

Thomas Mann: "If a man has a good cigar, then he's home safe."

"I don't understand," Hans Castorp said. "I don't understand how someone cannot be a smoker—why it's like robbing oneself of the best part of life, so to speak, or at least of an absolutely first-rate pleasure. When I wake up I look forward to being able to smoke all day, and when I eat, I look forward to it again, in fact I can honestly say that I actually only eat so that I can smoke, although that's an exaggeration, of course.

"But a day without tobacco, that would be absolutely insipid, a dull totally-wasted day. And if some morning I had to tell myself there's nothing left to smoke today, why I don't think I'd find the courage to get up. I swear I'd stay in bed. You see if a man has a cigar that burns well, and obviously it can't have any breaks or draw badly, that's really terribly annoying—what I'm saying is that if a man has a good cigar, then he's home safe, literally nothing can happen to him...

"Thank God, people smoke all over the world. There's nowhere you could possibly end up, as far as I know, where tobacco's unknown. Even polar explorers lay in a good supply of smokes to get them over their hardships...Let's assume for a minute that things would go miserably for me—-as long as I had my cigar, I'd carry on."
—*The Magic Mountain*, 1924
by Thomas Mann

EUGENE GANT PUFFS OUT

Eugene is a 13-year old paperboy: Assembled with three or four of the carriers in the lunch room, he learned to smoke: in the sweet blue air of Spring, as he sloped down to his route, he came to know the beauty of Lady Nicotine, the delectable wraith who coiled in his brain, left her poignant breath in his young nostrils, her sharp kiss upon his mouth.

And at 17, a stevedore: From time to time he thrust a handful of moist scrap tobacco into his mouth, and chewed with an air of relish. He spat hot, sizzling gobs upon the iron deck. God! thought he. This is man's work. Heave ho. There's a war on. He spat.
—*Look Homeward, Angel*, 1929
by Thomas Wolfe

BEYOND BASIC TRAINING

What's more important still is the issue of a double ration of smokes. Ten cigars, twenty cigarettes, and two quids of chew per man; now that is decent. I have exchanged my chewing tobacco with Katczinsky for his cigarettes which means I have forty altogether. That's enough for a day.
—*All Quiet on The Western Front*, 1929
by Erich Maria Remarque

STRIKING A POSE

She pointed to where Cheri stood, smoking a cigarette on the other side of the glass partition, his cigarette-holder clenched between his teeth, and his head tilted back to avoid the smoke. The three women looked at the young man who—-forehead at an angle, eyes half shut, feet together, motionless—-looked for all the wind like a winged figure hovering dreamily in the air.
—*Cheri*, 1929 by Colette

FOR WHOM IT MATTERS

"Do you have tobacco?"
Robert Jordan went over to the packs and opening one, felt inside an inner pocket and brought out of the flat boxes of Russian cigarettes he had gotten at Golz' headquarters. He ran his thumbnail around the edge of the box and, opening the lid, handed them to Pablo who

A Parisian bistro with caché.

took a half dozen. Pablo, holding them in one of his huge hands, picked one up and looked at it against the light. They were long narrow cigarettes with pasteboard cylinders for mouthpieces.
"Much air and little tobacco," he said.
—*For Whom the Bell Tolls*, 1940 by Ernest Hemingway

PRIMORDIAL SPARK

Fire is a gift to humans alone. Smoking cigarettes is as intimate as we can become with fire without immediate excruciation. Every smoker is an embodiment of Prometheus, stealing fire from the gods and bringing it back home. We smoke to capture the power of the sun, to pacify hell, to identify with the primordial spark, to feed on the marrow of the volcano. It's not the tobacco we're after but the fire. When we smoke we are performing a version of the fire dance, a ritual as ancient as lightning.

Does this mean that chainsmokers are religious fanatics? You must admit there's a similarity.

—*Still Life with Woodpecker*, 1980 by Tom Robbins

A Fast Burn

Playing at the Ford Theater the night Lincoln was shot was "Our American Cousin" by English-man Tom Kean. Even before unscripted tragedy brought down the curtain it was a most unfunny comedy. The climatic moment occurs when the careless Yankee of the title ignites a will with his cigar, accidentally incinerating a $400,000 inheritance.

GIVE ME A SPORTING CHANCE

Smoking is, if not my life, then at least my hobby. I love to smoke. Smoking is cool. Smoking is, as far as I'm concerned, the entire point of being an adult. I am well aware of the hazards of smoking. Smoking is not a healthful pastime, it is true. Smoking is indeed no bracing dip in the ocean, no strenuous series of calisthenics, no two laps around the reservoir. On the other hand, smoking has its advantage in the fact that it is a quiet pursuit. Smoking is, in effect, a quiet sport.

—*Social Studies*, 1980 by Fran Lebowitz

THIS MAN IS HAZARDOUS TO YOUR HEALTH

Nick Naylor had been called many things since becoming chief spokesman for the Academy of Tobacco Sciences, but until now no one had actually compared him to Satan. The conference speaker, himself the recipient of munificent government grants for his unyielding holy war against the industry that supplied the coughing remnant of fifty-five million American smokers with their cherished guilty pleasure, was now pointing at the image projected onto the wall of the cavernous hotel ballroom. There were no horns or tail; he had a normal haircut, but his skin was bright red as if he'd gone swimming in nuclear-reactor water; and the eyes—-the eyes were bright, alive, vibrantly pimply. The caption was done in the distinctive cigarette-pack typeface, "Hysterica Bold," they called it at the office. It said, Warning: Some People Will Say Anything to Sell Cigarettes."

—*Christopher Buckley*, 1994
Thank You For Smoking

Count Basie was one of the swing kings who pointed it out in smoke.

'Tobacco is a Musician

Technogamia, 1621

Smoky Notes

Jazz is the music of smoking, many aver. Ragged, edgy or mellow, riff by riff, its pulse and the smoker's heartbeat seem one. But not to me. Jazz is nervous, upsetting. But smoke soothes.

Any kind of music can be smoking music, and has been. Native American chants paced pipe puffs and some contemporary raps beat to the tobacco drag. The tobacco taverns of centuries past gave birth to smoking songs. And smoke music leapt out of the pub into classical art, as opera in Bizet's *Carmen*, and in a Bach song.

Erbauliche Gedanken eines Tobackrauchers (Edifying Thoughts of a Tobacco Smoker), found in *The Notebook of Mary Magdalena Bach*, is one of several songs the composer presented to his second wife, the mother of thirteen of his twenty children. Small wonder that Johann Sebastian had to duck out for a smoke—but one wonders how poor Mary

J.S. Bach, a Baroque Smoker

Leonard Bernstein's smoke trip was a musically distinguished journey.

took five.

The *Edifying* lyrics, by that international wordsmith, Anonymous, are of the ashes-to-ashes variety: both pipe and smoker are made of clay, both began as earth and will end as earth. Bach, the pipe smoker contemplates God as he puffs: Fame is but a fragrant smoke trip; pressing one's finger to burning tobacco is a coming attraction for hell. In short, smoking is a profoundly religious experience.

Did Bach smoke as he composed? Possibly. Leonard Bernstein certainly did; only with difficulty did he substitute a baton for a cigarette when he conducted. Even today, some ballet dancers dive for cigarettes during rehearsal breaks, although such pleasure is increasingly furtive. Nor are all opera singers abstemious. The Italian tenor, Luciano Pavarotti, admits that he has enjoyed cigars since he was little more than a *ragazzo*.

In this century, smoke rocks, smoke is bluesy, smoke is country. *Smoke! Smoke! Smoke that Cigarette!*, twanged by Tex Williams, sold a stunning two million platters in 1947. The Beatles have their hallowed tobacco lyrics; so do the Stones.

Tobacco Enemy Number One, James I, suffered smoke songs. He was in the audience for this surprise ditty to tobacco during a 1621 performance of the play, *Technogamia*.

> Tobacco is a musician
> And in a pipe delighteth.
> Tobacco is a lawyer,
> His pipes do love long cases.
> Tobacco is a physician
> Good for both sound and sickly.
> Tobacco is a traveler
> Comes from the Indies hither.
> It passed sea and land
> Before it came to my hand
> And this makes me sing.

The plaintive *Tobacco is an Indian Weed* dates to the same era.

> The Indian weed withered quite…
> That to ashes and dust return we must,
> Thus think and drink tobacco.

In 1719 the song was still current—versions persisted through the 19th century—but the lyrics had changed.

> Tobacco's but an Indian weed
> Grows green at morn, cut down at eve,
> It shows our decay, we are but clay.
> Think of this when you smoke tobacco.

An old sea chantey tells a tobacco-discovery tale:

> There were three jolly sailors bold
> Who sailed across the sea;
> They braved the storm and stood the gale
> And got to Virginee.
>
> Twas in the days of Good Queen Bess
> Or praps a bit before.
> Now hear how these soldiers bold
> Went cruising on the shore.
>
> A lurch to starboard, one to port,
> And forrad, boys, go we
> With a haul and a "Ho!" and that's your sort
> To find out tobac-kee.

Smoking songs were drinking songs and vice versa. Here's to *The Jolly Toper:*

> With my pipe in one hand and my jug in the other,
> I'll drink to my neighbors and friends.
> All care in a whiff of tobacco I'll smother,
> For the life I know shortly must end.
> What a hearty, what a hearty.
> He's gone.
> What a hearty good fellow.

One tobacco *Volkslieder* is virtually a planting prayer:

> Noble weed!
> that comforts life.
> And art with calumet pledges rife;
> Heaven grant thee sunshine and warm rain,
> And to thy planted health and gain.

Another old German smoking song hails the consolation of a pipe when other joys have flown:

> When love grows cool, thy fire still warms me,
> When friends are fled, thy presence charms me
> If thou art full, though purse be bare
> I smoke and cast away all care.

Lost love, of course, is the theme of *Smoke Gets in Your Eyes.* In the 1933 Kern/Harbach show stopper, smoke is the bittersweet memory that puts the tear in the eye. Some 90-odd versions of *Smoke Gets in Your Eyes* are still in the marketplace.

The 1930s were the heyday of smoke-along music. The 1936 hit, *These Foolish Things (Remind Me of You)*, recorded by both Billie Holiday and Benny Goodman brought us lyricist's Holt Marvell's image of a cigarette with

---

Hot Stuff

The average burning temperature of tobacco is 800 degrees F. Some cigarettes reach a heat double that. By contrast, the burning temperature of this book is Fahrenheit 451.

traces of lipstick. Bing Crosby's rendition of the equally-romantic *Two Cigarettes in the Dark* helped light him up as the crooner of choice for the very mellow.

Big Tobacco guaranteed the presence of the smoking lyric in America's living rooms, sponsoring such radio shows as *The Chesterfield Supper Club*, whose theme song was *Smoke Dreams*. Love, heartbreak and cigarette smoke are inseparable in the song, written by John Klenner, Lloyd Schaffer and Ted Steele:

> I try to forget with each cigarette...
> Still smoke rings bring smoke dreams.

Big Bands and their vocalists specialized in gloriously smoky music. Casa Loma's theme was *Smoke Rings*, one of the more alluring songs of phantom love:

> Where do they go—the Smoke Rings
> I blow each night?
> What do they do—these circles of
> blue and white? Oh.

Singer k.d. lang puffs new life into the smoky blues.

> Why do they seem to picture a dream
> of love?

Ned Washington's *Smoke Rings* lyrics are revived with sultry perfection in k.d. lang's 1997 CD, Drag. Her voice is awesomely smoky although she doesn't smoke. On the same CD,

"Clearing The Smoke," 1994
by Margaret Dolinsky

This is my last cigarette.

The smoke-filled cabaret is a signature of the past in some cities, even if its lovelorn enchantments linger. Lived there ever a piano player or a melancholy barman who didn't smoke? Billy Joel picked up this theme in *The Piano Man*, where the bartender is quick to light patrons' cigarettes, even if he is dreaming of life elsewhere. *Piano Man* was recorded in 1974. Most bartenders are still handy with a match, although there are a sour few—God help us!—who will add, "That stuff will kill you." As if we didn't know. As if we hadn't always known.

Smoke! Smoke! Smoke that Cigarette, the Tex Williams 1947 mega-hit, written with Merle Travis, was a tongue-in-cheek reaction

she interprets *My Last Cigarette*, the achievement of Boo Hewerdine, Gary Clark, Neil MacCall and Dizzy Heights. In this twist, the equation is between unkind lovers and cigarettes:

> I have a habit I've been trying to lose.
> Everyone thinks they know what
> they want.
> Sometimes a drug chooses you...
> It's my last cigarette.

to the smoke-filled world of a half century ago. More bluegrass rap than tune, this is the lead-in:

> Now I'm a feller with a heart of gold,
> With the ways of a gentleman I've been told,
> The kind of feller who wouldn't harm a flea.
> But if me and a certain character met—
> The guy that invented the cigarette
> I'd murder that son-of-a-gun in the first degree.

The song goes on to tell two stories-- one where Tex can't cash in on a poker hand, and one where he can't cash in on his plan to kiss a girl. The obstacle? At the critical moment the other party lights up.

The Western Caravan sings the chorus:

> Smoke, smoke, smoke that cigarette.
> Puff, puff, puff it but if you smoke yourself death...

Tex picks up:

> Tell St. Peter at the Golden Gate
> That you hates
> To make him wait
> But you gotta have another cigarette.

This oddball number was recorded again in 1976 by Commander Cody and His Lost Planet Airmen.

Now I have a slight special interest in the commander because in his real life—when he had a real life—he was my Bay Shore High School classmate, George Frayne. Back then, he was a middle class arrow, a broad-shouldered football player who dated the blond cheerleaders. But he could hit a mean jazz piano. The bottom-line question: did this piano player smoke?

The answer: very little. The rules were different for halfbacks—they couldn't smoke upwind, downwind, or in gossip distance of the coach.

He shot off to the University of Michigan and transformed into the leader of a pop bar band with a sound of its own: country-trucker rock. Commander Cody wrote a few songs (good ones) and also updated *Smoke! Smoke! Smoke that Cigarette!* He made that kiss-girl into a long tease:

> She said, "Cody, excuse me, please,
> But I only get off on 100 millimeter cigarettes."

The commander's next flourish: "Now you all don't know this is a message tune but it is. So if any of you fools are trying to quit, I ain't gonna criticize you a bit."

The finish flips us back to the pearly gates. "I'm talkin' about the last, the final, the ultimate cigarette," Cody thunders. Another 1947

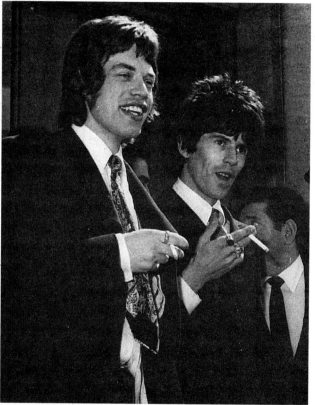

Mick Jagger and Keith Richard get "satisfaction" in 1967, after being charged in London court with drug possession.

cigarette biggie warrants notice: Tim Spencer's *Cigarettes, Whusky and Wild, Wild Women*, recorded by the Sons of the Pioneers. It's one in the double-standard vice chain, linking tobacco, alcohol and sexy women.

> I met with a gal, and we went on a spree.
> She taught me to smoke and drink whiskey...
> Cigarettes and whusky and wild, wild women.
> They'll drive you crazy, they'll drive you insane.

Fast forward to classic rock. Cigarettes are as much part of the scene as screaming fans. Smoking is an antidote to teen boredom. In 1965, the Statler Brothers sang it loud and clear in Lewis Dewitt's *Flowers on the Wall*:

> I'm having lots of fun...
> Smoking cigarettes and watching
> Captain Kangaroo.
> Now don't tell me I have nothing
> to do.

That same year, The Rolling Stones recorded *Satisfaction*, whose lyrics imply that cigarette smoking is a prerequisite for friendship, even manhood.

The Beatles were "so nervous" in 1965, when they were about to be made Members of the British Empire, that right before meeting the queen "we went into the toilet and had a cigarette. In those days we all smoked," remembered George Harrison, thirty years later.

In 1967, John Lennon sang of smoking on a double-decker bus in *A Day in the Life*. Much in the news that year, and the next, 1968, was the word that cigarettes were bad for your health. In *I'm So Tired*, John lit up another cigarette while swearing at Sir Walter Raleigh.

Long-distance American busses, as well as the top tier of British city busses, are two of the many places you used to be able to smoke. Cigarettes were integral to Paul Simon's 1972 *America*, where the quest starts with the purchase of a pack.

Aside from Lennon's dig at Raleigh, the U.S. Surgeon General's Report had about as much immediate impact on pop lyrics as it had on movies---nearly none. Nineteen seventy-four heard Brownsville Station tell the gold-record truth in *Smokin' in the Boys Room* (by Michael Koda and Michael Lutz):

> Smokin' in the Boys Room,
> Yes indeed, I was smokin' in the
> Boys Room.
> Now teacher, don't you fill me up
> with your rules
> For everybody knows that smokin'
> ain't allowed in school.

This truth may not be eternal, but it

Dear Old Mummy

Did ancient Egyptians smoke tobacco? Or eat it? It's a stunning possibility.

In 1992, a team of scientists at the Institut for Anthroplogie and Humangenetik in Munich, Germany, led by Svetla Balabanova, decided to take the wraps off the former lifestyles of mummies. Testing the hair, bone and soft tissue of seven mummies, dating from 1070 BC to 395 AD, they found prodigious residues of nicotine.

No doubt about it---the Egyptian royals and priests enjoyed the high life. But where did they get the tobacco?

Tobacco is a New World plant, unknown—or so we thought—on the rest of the planet before Columbus. New theories abound. Perhaps the ancient Egyptians, the techies of their day, traded across more oceans than Europeans dreamed of. They may have traversed the South Atlantic or encountered South American sailors, and so came by tobacco; they somehow met the Chinese, or people who knew people who knew them, and so obtained the scraps of silk which have also been found in mummy hairdos. But there is little evidence other than the presence of nicotine (and silk) in ancient Egyptian culture to suggest heroic martime feats.

British toxicologists, dazzled by the Munich discovery, set to work on some of the old boys and girls resting in the Egyptology section of the Manchester Museum. They, too, found enough nicotine to mark those old mummies as chain smokers. Moving ever backward, Balabanova and her colleagues tackled the bones of some Neolithic citizens of Turkey and Jordan. Those prehistoric skeletons also gave up nicotine ghosts.

Did everyone who was anyone smoke once upon a time? Was taking tobacco part of long-ago medical or religious experience in the Near East?

Debate rages. Some scientists suspect specimen contamination, thinking of the 19th-Century pipe smokers who once cradled newly-discovered mummies or handled old bones. The counter-argument goes that the nicotine in hair samples appears to have been metabolized, indicating that the antique Egyptians gobbled up tobacco, or inhaled its smoke, while they were still spry.

Others wonder if the nicotine might have been derived from another plant with trace levels of the substance, such as European belladonna, once prized as a medicine, cosmetic and poison. Apparently, it would require an incredible level of belladonna addiction—or some very talented and frequent poisonings—to come up looking like a mummy.

Just where the people who became mummies found nicotine is not likely to be answered soon. In the meantime, your opinion is as good as mine. Personally, I can't believe that any society as sophisticated as that of the Pharaohs could "discover" the golden tobacco plant and not seed it on home soil. However the dear old mummies got their nicotine fix, they never saw tobacco bloom.

was going strong in 1985 when Motley Crew revived the tune, and it holds today. Cigarettes sign defiance. Contempo pop smokes in a decidedly old-fashioned way. The 1995 Oasis CD, *Definitely, Maybe*, includes the Gallagher brothers' *Cigarettes and Alcohol*, in which the bored young hero declares that the title substances are all he needs for contentment.

There are some new blues entries on the addiction front. *Crazy Baby*, written by Rodney Crowell, notes:

> And your hands are shaking
> something awful
> As you light your twenty-seventh
> cigarette.

It's not the cigarette that's giving Baby the shakes, but Crowell picked the magic number as far as smoking goes. The average smoker needs exactly twenty-seven cigarettes per day to maintain a happy nicotine level.

Among Drag's love-dependency numbers is David Barbe's *Your Smoke Screen*. Needing love so badly you make a bad pick is a familiar poison. The refrain:

> I should have seen through your
> smoke screen.

One of Jerry Garcia's final acts of grace was his rendition of the Otis Redding-made-famous ballad, *Cigarettes and Coffee* for the *Smoke* soundtrack:

> It's early in the morning,
> About a quarter to three.
> I'm sitting here talking with my baby
> Over cigarettes and coffee.

This classic was written by Jerry Butler, Eddie Thomas and Jay Walker in an era when the smoking man and smoking woman held their heads high unless they were sighing over love gone by, which they almost always were.

Nostalgia is the ordinary vice of old and young alike—listen to a fourteen-year-old mourn yesterday. Between what passes for new and what's golden oldie, the sound waves are a smoking zone.

'There's no sweeter tobacco comes from Virginia and no better brand than The Three Castles.'

William Thackeray
The Virginians, 1859

Ad-ing It Up

America's first citizens grew their own. If there was a rumor that one trail mix was superior to another, it was not recorded for posterity. But once Europeans entered the scene, tobacco and advertising nearly became synonyms. Not long after Pocahontas' English sojourn to promote her husband's import, newspaper ads and fliers announced the arrival to market of new tobacco shipments.

Snuff was the great boon to the nascent craft of advertising because tobacconists created and packaged their own blends, allowing the claim of an extraordinary product.

By the 1700s, big-city tobacco merchants had enough promotion tricks up their sleeves to make a Saatchi cry. Among them were riddles placed in newspapers, the answers available only at the advertiser's shop.

A 1714 tobacco box sported this verse:

> Were I not confined in narrow space
> The virtues of this wondrous Herb
> I'd trace...
> How it Collects the thoughts and serves instead
> Of biting Nails or harrowing up the Head.
> But this task I leave to Future Rimes
> And Abler Poets born in better Times.

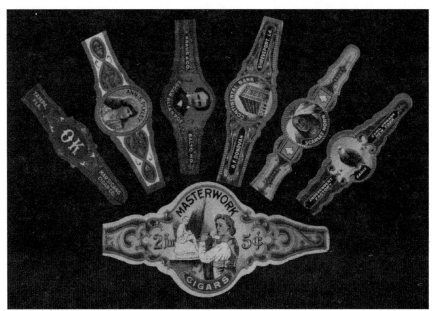

Tobacconists handed out trade cards, enlivened by verse or jokes. The practice persisted into the early 1800s; this bit of tobacconist poetry dates from then:

> I owe to smoking, more or less
> Through life the whole of my success
> With my cigars I'm sage and wise—-
> Without I'm as dull as cloudy skies.

The advertised virtues of a brand almost always promised its consumer a high time. Typical was an 18th-century ad in the London *Spectator* which declared that Angelick snuff "revives the spirits."

By the mid 1700s, tobacconists used papers printed with brand names to wrap chew, snuff powder and cigars.

From the earliest days of tobacco import, a critical selling point was place of origin, the more exotic the better. London tobacconists invented the cigar-store Indian to advertise American tobacco. The guardian Turk, signaling Anatolian tobacco, was a popular figure for both European and American merchants. A kilted Highlander pledged Scotch-style snuff. Crinkly-haired figures might suggest products from Jamaica or Havana, but they were also used to invoke Virginia.

Cigar bands and boxes were more important. The bands, which originated in Cuba as a quality guarantee, were copied and even counterfeited to dress cigars made elsewhere. Packaging mattered. Beautiful cigar box labels hinted at paradises far beyond the enjoyment that a good cigar might bring its smoker. Seals testified to the integrity of the container's treasure.

Some cigar boxes also included coupons to be hoarded, then cashed in for merchandise including, in one instance, a home. The contemporary "Joe Camel" coupon, shocking to so many, typifies one of Tobaccoland's longest-running gimmicks.

The big collectibles on the chew front were tin tobacco tags dropped in the bags. Retrieved, they were redeemable for other merchandise.

A 19th-century pipe-tobacco manufacturer went a step further, promoting sales by randomly dropping paper

Tin tobacco tags.

cash in its product bags. The stratagem worked.

John Green, who trademarked his Bull Durham smoking blend in the 1870s, is sometimes credited with establishing the parameters of modern tobacco advertising. The North Carolina entrepreneur authorized billboards, paid bonuses to aggressive middlemen, solicited endorsements from such distant stars as Alfred Lord Tennyson. Bull Durham's success— its nine hundred hands made it the biggest tobacco plant of its day— invited imitators, Sitting Bull chief among them.

Nineteenth century tobacco-manufacturers mounted professional events. Gold Coin Cigar backed a national walking contest, raising fewer eyebrows than did Virginia Slim's sponsorship of

women's tennis a century later. Cigarette makers also hit upon prominent athletes to promote their wares.

Baseball cards weren't born in bubble gum packs as the uninitiated believe. Early this century, they were inserted in cigarette (and candy) packages, successors of tobacco silk squares and lithographed cards, which had been current since the 1880s. Series on silks or cardboard, which stiffened paper wrapping, covered a riot of subjects: birds, flowers, household cleaning tips, battle scenes, Indian chiefs, to name a few. Pipe tobacco, such as Prince Albert, was also sold with trophy cards. Teasing the collecting instinct of man, woman or child to win brand loyalty is an old idea.

Tobacco brand names are a hoot; their implications could stuff a thousand doctoral dissertations.

Most cigars long have paraded proudly under Spanish names.

Plug has sailed under nautical banners and other flags: Raleigh invoked history; Scalping Knife or Battle Ax fired up memory of frontier war; Lucky Strike was born in Gold Rush days. Names could be candid: Cheap John, Gold Brick, Old Hat. Some chew labels were blunt: Jaw Bone, Mule's Ear; others, like Honey Suckle, were coy.

Cigarettes went in for uppity labels. Vanity Fair and Entre Nous were named to snare swells. A hundred years ago there were dozens of cigarette brands and even more varieties of brand-named shred for the roll-your-own majority. In a salaam to Constantinople, the geographically-challenged stamped exotic names on blends: Fatima, Hassan, Egyptian Deities, Egyptian King, Murad. Other popular labels included Pet, and the penny-apiece Sweet Caporal or "Sweet Caps," made in New York City by the Kinney company from a mix of Piedmont Bright dashed with spicier Louisiana and Turkish tobaccos.

English brands were schizophrenic. The lure of the mysterious East beckoned the "upstairs" crowd while the claim of "Virginia tobacco" hooked the hordes below. Cigarette purchase was made easier for all classes, thanks to a Londoner's invention of the vending machine shortly after 1900.

American brand names turned Anglophile as the 20th century puffed on: Marlboro, Kent, Parliament, Viceroy, Chesterfield, Pall Mall. There are prominent exceptions to the Brit-clique, of course.

Paradoxically, the image pushed for even the English-sounding brands is American red, white and blue.

The rest of the world, save proud France, just wants to smoke what Americans smoke and be who they think Americans are. Moscow has turned from Red Star to Marlboro, but George Bush's trade gang failed to make China safe for Marlboro (or Marlboro safe for the Chinese). Most of China's three hundred and fifty million smokers have access only to Panda and its domestic kin.

LEADERS OF THE PACKS

Twenty-one year old James Bonsack changed the tobacco world and upped the advertising ante by inventing the cigarette rolling machine. Before 1884, cigarette advertising was limited by the amount of product human hands could turn out. Bonsack's two-thousand pound critter spurted forth two hundred cigarettes per minute, matching the productivity of fifty human beings.

James Buchanan "Buck" Duke, all grown-up and head of American Tobacco in Durham, N.C., grabbed the rights to the Bonsack machine. He was soon manufacturing more cigarettes than chew. Duke cranked up the grand old tobacco promotion traditions of trading cards and coupons. Seventy-five coupons would "buy" a picture book of famous actors or one of "sporting girls." A lithographed-card series enclosed with Cameo cigarettes was entitled "Histories of Poor Boys Who Became Rich and Famous." With

Gift redemption coupons are an old hook.

accuracy-be-damned fervor, it included the likes of George Washington and Thomas Jefferson.

Duke was also big on newspaper advertising.

In 1889, Buck Duke's factories produced two million cigarettes per day—his leading brand was Duke of Durham—which sold at prices that torpedoed the competition. With his immense profits, Duke gobbled up other tobacco ventures.

Richard Joshua Reynolds had ridden his horse into Winston Salem, N.C. in 1874 looking for land on which to build a tobacco factory. He bought a large plot from the Moravian Church for $388, the first of many sharp deals. Reynolds would eventually outdo Duke, but in 1888, he must have been nodding out over his mountains of plug. He did not make the leap to cigarettes for decades. Reynolds' big brand, introduced in 1913, was Camel—the package illustration a likeness of a decrepit Barnum and Bailey circus specimen known as Old Joe.

P.T. Barnum, whose motto was, "There's a sucker born every minute," had campaigned against tobacco. Whether R.J. appreciated the irony when he adopted Barnum's camel is unknown.

The launch of Camel was a triumph. Reynolds bought out and buried the small brand, Red Kamel, to guarantee his exclusive right to the animal of

Trophy cards are almost as old as the tobacco hillocks.

Arabian nights. Pyramids and palm trees were added to the package design. Reynolds went big time with his "Turkish and domestic blend," hiring the N. W. Ayer advertising agency to give its all.

Camel was introduced at a discount price and went national immediately, with ads in *The Saturday Evening Post*. The launch cleverly marked the dromedary's progress, starting with "The Camels are Coming" and ending with "Camels leave satisfied smokers." Ad copy mocked the give-away legacy, trumpeting: "Fine Tobacco, No Premiums."

Right before World War I, when cigarette advertising had yet to become a magazine mainstay, a few journals followed the lead of Henry Ford and Lucy Gaston and ran articles warning of smoking hazards. The *Literary Digest* published "Killing the Cigarette Habit" in 1913. The next year a provocative piece headlined, "What's the Matter with My Pulse?" appeared in *Cleveland Ladies Home Journal*.

By 1921, Camel owned half the American cigarette market. That same year, the R J. Reynolds company snatched an offhand remark made to one of its ad men—"I'd walk a mile for a Camel"— and made it the brand's slogan. Years later, Ayer came up with the "T-Zone" and "More doctors prefer Camels."

All the majors relied on celebrity endorsements. Fred Astaire declared, "I made my own thirty-day mildness test. It's Camels for me from now on."

Henry Fonda avowed, "My voice is important in my career. I smoke Camels because they're mild."

Meanwhile, Buck Duke's empire had been trust-busted and he had parachuted into British American Tobacco. Duke's successor at American Tobacco, George Washington Hill, resurrected the pipe-tobacco label, Lucky Strike, for a cigarette meant to upstage Camel.

As some men prone to violence possess two "Y" genes, Hill seemed to have been born with a secret chromosome coded for advertising acumen. He seized on the subliminal notion that cigarettes kept you trim, and coined "Reach for a Lucky instead of a sweet." Sales of Luckies, wrapped in money-green, took off.

A later Lucky Strike slogan—"It's

toasted"—was high hokum, claiming distinction where none existed. All Bright tobacco is heat-cured. Equally silly, but with a certain zing was "L.S.M.F.T.— Lucky Strike Means Fine Tobacco." Far snappier was "Be Happy, Go Lucky!" American Tobacco was so cheered by its wit, it erected a Lucky Strike Pavilion at the 1939 World's Fair.

Bigger than Lucky was Liggett & Myers' Chesterfield. The St. Louis-based Liggett had been the prince of chew when Duke's kingdom swallowed it. It took some time for Liggett & Myers to try cigarettes after Sherman Act intervention set it free.

Liggett considered it daring in 1927 to tiptoe up to women with Chesterfield ads featuring a man smoking while the lady in his life begged, "Blow some my way." Chesterfield

advertising evolved to featuring a female tennis player with "flash."

The brand underwrote *Chesterfield Supper Club*, the national radio music program which launched low-key, if not off-key, Perry Como. Listeners from coast to coast also had dubious benefit of Liggett's visit to the alphabet for a catch-phrase: "A.B.C. Always Buy Chesterfield." The all-purpose, if derivative, motto, "They Satisfy," worked best. Whatever Chesterfield lacked in slogan pizzazz, it made up for in print visuals.

Chesterfield and Camel each averaged twenty-seven percent of the American market through the 1930s, with Lucky Strike five points behind.

Moderately successful brands of the period included Old Gold, produced by P.

"Old Gold" once danced for the New York Heart Fund.

Lorillard, once a New Jersey snuff maker, and coupon-laden Raleigh, made by Brown and Williamson (B & W) in Louisville, Ky.

B & W introduced mentholated Kool and its penguin in the '30s, after buying out Spud, the first cigarette doused in peppermint extract. Spud had been a dud, showing a pitfall of naming a brand after its creator, but mentholated Kool, with its refreshing name and snazzy penguin, was a comer. In the same decade, B &W challenged Parliament with Viceroy, whose cellulose filter became the industry standard.

The Philip Morris pageboy singing

"Any KOOLS Today?"

ORDER something easier and cooler for your throat today. Order KOOLS. There's just enough mild menthol to keep your throat easy and your mouth sweet... but the grand tobacco flavor is there in every refreshing puff. Cork tips protect your lips. Each pack carries a valuable B & W coupon worth saving for handsome premiums. Summer's a-comin'; take the parch out of smoking: switch to KOOLS.

RALEIGH CIGARETTES...NOW AT POPULAR PRICES...ALSO CARRY B & W COUPONS

out, "Call for Philip Morris" debuted on the company's weekly NBC radio show in 1933. The original brand had been an expensive Turkish brew formulated in London. The new blend, paged by Johnny Roventini, was an American mix. Pint-sized Roventini had actually been working as a pageboy in New York hotels when he was summoned to his major role in life. His Brooklyn-accented pitch worked wonders.

Some 1930s brands didn't make it, among them Barking Dog and O-nic-o, which claimed to contain less than one percent nicotine. However, the Depression sparked sales of tobacco for roll-your-owns, and small vendors sold branded cigarettes as "loosies," one at a time.

The big three, Chesterfield, Camel and Lucky, rode high on masterful wartime ads but were toppled at the end of the '40s by American Tobacco's long, slim Pall Mall, which successfully drew in legions of young women, including the glamorous blonde who became my mom. The Pall Mall motto was: "Wherever particular people congregate." The most usual place to meet my mother—outside her suburban home—was at the discount supermarket, Korvettes. Pall Mall ads also promised that its smoke didn't irritate the throat.

R.J.R.'s Winston and Lorillard's Kent, (first marketed in 1952) were the blockbuster filters of the late '50s and '60s.

Kent assuaged smokers' health concerns by boasting a "micronite" filter which blocked "heat." Some smokers felt it also blocked taste. More significantly, the fil-

ter contained asbestos. More than forty years later, that filter went on trial in California and was convicted of promoting Prof. Milton Horowitz's asbestos-related tumor, mesothelioma. Lorillard and the filter maker, Holingsworth and Vose, were ordered to ante up $2 million worth of damages.

Winston, named after Reynolds' home but suitably English in sound, was introduced on television in 1954. The simple if ungrammatical "Winston tastes good like a cigarette should" got at Americans where they lived: in front of the box. Within five years it was America's best-selling filter cigarette.

An R. J. Reynolds analyst, looking back at the brand's success with younger Americans, commented, "Winston let Kent and Viceroy sell the benefits of filters so it wouldn't appear a sissy brand."

But a more macho competitor was

The Marlboro Man in 1944, before his sex-change operation.

mounting up.

The Marlboro Man used to be a woman. And she was a cosmopolitan type—or that's how Philip Morris once angled Marlboro advertising. Approaching middle age, the Marlboro Woman underwent a sex-change operation, emerging with new lifestyle options.

The Leo Burnett ad agency hung several occupations on the Marlboro figure's broad shoulders. "Where there's a man, there's a Marlboro," was the new battle cry. Green Bay halfback Paul Hornung, 1961 National Football League "Player of the Year," was one of the last major athletes to celebrate smoking openly: "Paul's a Marlboro Man all the way," ads boasted. As the decade smoked on, the Philip Morris cowboy galloped past his alter-egos.

Winston was knocked from the saddle. Marlboro went from roping eight and a half percent of the 18-year-old (new smoker) market in 1964 to lassoing over half of it in 1983.

In the '70s and early '80s, health-concerned smokers traded in unfiltered loyalties for seemingly-safer brands.

My mother, no longer a trendette, switched to white-tipped Kent and then to Benson and Hedges 100mm (remember the terrific TV commercial where the elevator door closes on that ultra-long cigarette?) and Salem Lights. My impression is she smoked more of these milder choices. Studies show that people draw as deep, and smoke as many sticks as needed to maintain customary nicotine level. Whatever—the switch didn't help. Her fatal vascular disease was probably smoking-related.

But I'm not blaming her or the advertisers or anybody else. I like much of the advertising—some enthralls me almost as much as old movies do.

OUT OF THE MOUTHS OF BABES

Children were viewed as miniature adults for much of smoking history and so out of the mouths of babes drooled nicotine. Tobacco was pasted on infants' gums to quiet teething pain.

Young children in both Europe and America shared adults' pipes. The practice was not so common in upper-class homes, where the preferred form of tobacco was snuff. Perhaps youngsters had the good sense to resist an experience that led to sneezing seizures, or maybe they were judged incapable of mastering the elaborate etiquette of snuffing.

The pipe, particularly the breakfast pipe, however, was a regular part of many children's lives, often as a substitute for food. Very small boys and girls clutched parents with one hand and smoked with the other. In the United States, many boys chewed tobacco.

Nineteenth-century social custom diverged from the past. Among the tenets applying to the Victorian well-to-do were that men were the intellectual superiors but moral inferiors of women, that children should not bear the responsibilities and privileges of adults, and that piano legs and lamb chops were so sexually-alluring that they should not be seen undressed. I'll leave the discussion of musical instruments and meat to finer minds.

The brunt of Queen Victoria's tobacco position fell upon women and children. The main rap against tobacco in the 19th and early 20th centuries was that it was immoral. Remember, the savage weed. It was also held to undermine the good work habits, including those of children, required to keep the Industrial Age grinding.

Ministers sprung up preaching an anti-tobacco gospel. One of the foremost was the Reverend George Trask of Boston who penned his version of "Just Say No" in 1852. The title of his abstinence address was *Uncle Toby's Anti-Tobacco Advice to His Nephew Billy Bruce*.

Psuedo-medicalist Orson Fowler had a special message for young tobacco eaters: "You, who would be pure in your love instinct, cast this sensualizing fire from you."

In 1872, the crusading *Anti Tobacco Journal* appeared on the scene.

A segment of entrepreneurial America and its advertising agents saw opportunity here. Nostrums came to market —No-To-Bac, Narcoti-cure—promising salvation from the tobacco habit.

Seventy-five years ago tobacco-cures were promised that we still don't have.

I guess they didn't work.

Mirroring the misogyny of the era, self-appointed authorities were not sure which was worse, boys smoking or women adopting cigarettes. Dr. R. T. Thrall railed understandably at three-year-olds smoking on New York streets but was equally perturbed by "ladies of this refined and fashion-forming metropolis...smoking Tobacco through a weaker and more feminine article...the cigarette."

The tobacco industry went its merry

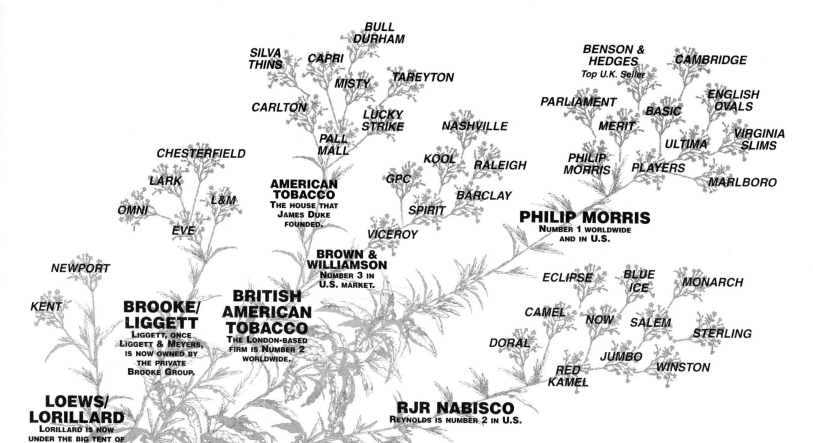

Who's Who in Cigarettes

Here are the better known among the hundreds of brands available in the United States.

way devising come-ons aimed at segments of population where tobacco buying might be increased.

Women were offered perfumed cigarettes through the 1920s. Candy-flavored eating tobaccos were merchandised for those with juvenile tastes.

Recruiting young smokers is not new, but in recent years it has been desperate.

The percentage of the population that smokes has declined steeply from the halcyon tobacco days of the 1970s, and neither hard-core quitters nor the dead are likely to resume. In the United States, only one in three adults now smokes, but in the last few years, the percentage of smokers has risen slightly. The fastest-growing market segment is the young female.

A slew of new U.S. government rules limits youth-directed advertising. Some of the regs make sense, but one—— requiring that sales clerks demand photo identification from tobacco purchasers who appear to be "under 27"—-is an over-the-top response to Corporate Tobacco's deep dives for child subscribers. In teen speak, it's "random." In the jargon of constitutional law, it's a violation of the right to privacy.

Cynics say that most cigarettes taste alike, which is why advertising matters so much. Tests of blindfolded smokers indicate the cynics are right. But I say there's room for doubt. Maybe some smokers need packaging cues to distinguish one smoke from another. But the smoker is a paradox. Someone else's brand usually disappoints, but in a state of nicotine deprivation, smokers never meet a brand they don't like.

The trick for advertisers, then, has been to dangle their product in front of those too new to smoking to be brand-committed. They seek the young. Teens nick cigarettes or

> ### Smoke of the Hindu Gods?
>
> *Slender "beedies," hand-rolled in dark tendu leaves, and sometimes flavored with strawberry or clove, are the cheap, Everyman smoke of India. But among American and European students they've picked up a rep as aphrodisiacs, with price tags to match. Hot brands come with deified names, such as Kalia and Mangalore Ganesh. If they spark illusions among their worshippers, it's likely because a beediecontains from twice to eight times as much nicotine as an American cigarette.*

What Price, Happiness?

When Woodrow Wilson's vice-president, Thomas Marshall, complained, "What this country needs is a good five-cent cigar," there still were nickel stogies. Evidently, they weren't up to Marshall's standards.

The cheap cigar problem had bothered the writer William Thackeray half a century earlier, during a visit to Ireland. "Throughout the town of Galway you cannot get a cigar that costs more than twopence," he grumbled. "Londoners may imagine the strangeness and remoteness of the place."

Robert Louis Stevenson, who made his South Pacific island paradise complete by rolling his own cigarettes, found tobacco dear but didn't begrudge the expense. "There are not many books which are worth the price of a pound of tobacco," he registered in "An Apology for Idlers."

In today's America one can buy a rotgut cigar with pocket change or, as reported by The New York Times, an illegal Cuban for $28. I've had the latter offered to me for more. The international tax-free price of Cohiba, the cigar originally produced for the delectation of Fidel Castro, is about $30. Some cigar pros aver that great Cubans aren't better than top Dominicans, only scarcer.

The 1890s were the heyday of the nickel pack of ten cigarettes. Then the Spanish-American war tax sent the price to six cents. The penny cigarette flourished after the turn of the century. The new Philip Morris debuted in 1932, selling at fifteen cents for a twenty-cigarette pack.

The U.S. wartime Office of Price Administration fixed cigarette prices in 1942: twenty cents for a standard pack (with some cigarettes selling cheaper). Two decades later, the twin-dime pack was still available in the South. In other regions, cigarettes cost a few cents more. The steepest U.S. price increases occurred between 1981 and 1986, as the number of smokers started to decline. The average cost per-pack rose thirty-six cents, only one-third of that going to taxes.

When tobacco made its first bow in England, its value was literally measured in silver, pound for pound the substances were of equal worth. Then taxes were added. As supply increased, price plummeted. In 1675, Virginia growers received tuppence for a pound of tobacco that retailed at nine pence.

Tobacco was currency in the colonies. It paid for the passage to Jamestown of twelve English maidens exported as brides for a dozen lusty bachelors.

It was not uncommon for tobacco to be part of wages in America and elsewhere. Peter Kolb, a Dutch visitor to South Africa in the early 1700s, was left as discomfited by a tobacco shortage as were the African laborers who expected it as part of their pay. Kolb saw how the locals coped and copied them, putting elephant dung in his pipe. He thought, "the scent and flavor of it were pretty much like those of tobacco." But without the nicotine.

sponge them or buy them where they can't be fussy—-brand loyalty is up for grabs.

In Big Tobacco jargon, the search has been for "replacement smokers."

Their current documents are careful to label their target group "18 to 21 years old," although they know as well as you and I that most smokers start younger. They also know that ninety percent of people who don't begin smoking until they're 21 will quit within five years.

The most successful and expensive tobacco campaign of all time is that for Marlboro Country, which has oldsters coming back for more while it pulls in youngsters and keeps an impressive sixty percent of them loyal to the brand. Philip Morris has been spending about $100 million annually to promote Marlboro.

Philip Morris, whose brands as a whole corralled half of the U.S. market for the first time in 1997, also underwrites significant cultural events, from arts performances to constitutional debate. When you're Philip Morris, it's expensive to keep up your good name.

In the past, the corporation has been a genius at acquiring unorthodox product display.

In *Superman 2* (1980), Lois Lane chain-smokes Marlboro, and a shiny red and white Marlboro delivery truck is stage center in the climatic Superman-saves-Gotham scene. When Hollywood really smoked—cigarettes, that is—labels were removed or hidden. Now spin-offs, merchandising and extra revenue streams can green-light a picture. However, *tobacco* product placement may soon be against the law.

Marlboro's wild success has provoked attacks. The Houston, Tex. branch of Doctors Ought to Care commissioned a "Barfboro" poster, now part of a Smithsonian collection. Marlboro Country ads ride the roofs of some New York City taxis. So does a counterblaste "Cancer Country" that puts a skull under the

white hat of the cowboy.

Boldest of the assaults comes courtesy of California, where even bars must soon be smoke free. In a thirty-second TV spot, devised by Asher/Gould Advertising, an ersatz Marlboro Man prods children into a corral as an unseen voice intones, "Once they get you where they want you, they got you for good."

In England, the Marlboro Man is an outlaw. All advertising images of vibrant, healthy smokers are banned. Similar restrictions are coming in the United States. Even toon figures are headed for the ashcan.

Anti-smoking forces haven't been the only students of Marlboro's achievement. Its competitors secretly sat at the master's feet.

An internal R.J.R. Reynolds memo, written a few years back and analyzing the success of Marlboro with "replacement smokers," was so envious it might have been written in green ink. The document stressed that "young smokers want to prove their independence and maturity through their smoking."

In 1996, Reynolds spent $30 million—down from $75 million—pushing Camels. Young Joe Camel, born in 1988, is as flashy a dromedary as ever humped the desert. It is almost sad to realize how likely he is to die young.

Actually, there will be life of a sort for Joe, the Marlboro Man and other tobacco heroes. They will live on in more tolerant nations, admired for their American spirits although exiled from their homeland.

Does Joe deserve more honor in the land of his birth? Arguably, Reynolds nurtured him to seduce youth. The Joe-toon C notes in each pack, redeemable for merchandise, may be an antique ploy but they made the government hit list, along with giveaway baseball caps and T-shirts, because of their appeal to the young.

The irony is that Joe was not as much of a success as most Camel-bashers believe. The unfettered Marlboro cowboy, roaming the plains under the great blue sky is the person teenagers admire. So what if the Marlboro man has sun-and-wind lines on his face; it doesn't prove tobacco ages the skin, it shows what a great life this master of his own destiny leads. Joe Camel may have been the darling of five-year olds, but when teens belly up to that cigarette counter, they don't want to be mistaken for cartoons.

Reynolds' recent revival of Red Kamel is an exercise in nostalgia and possibly a tacit admission that Joe is just too cute to live.

Cigarette makers have not been the only tobacco merchants fishing in the kiddy pool. Special promotions for dip and chew have also targeted youth. The United States Tobacco Company is the biggest purveyor of smokeless tobacco.

Meanwhile, cigarette producers have been boning up on how to sell their wares in muted ways. Quiet marketing techniques—some of which can flourish in the new age—have advanced the discount "generics," such as GPC and Basic. Corporate Tobacco has found these little-advertised brands very profitable.

Boutique brands are also hot. Selectively-distributed boutiques, for instance Reynolds' Blue Ice, are bets on highly-stylized packaging and awesome names to win over younger smokers.

Boutique brands, an echo of the past, bespeak the future, or a profitable part of it, if vendors aren't required to hide their merchandise. The boutique needn't have a distinctive taste just a fresh cover look and a beguiling name. Without big advertising, such cigarettes can be priced as high or higher than name brands. These "personalized" packages can be designed to accompany any lifestyle and be styled for smokers of any age.

Oddly, Big Tobacco still puts great faith in separate feminine thought processes. Philip Morris' Virginia Slims traded in "You've

come a long way, baby" for "It's a woman thing." The counterattack: "Virginia Slime—it's a cancer thing," underscores a drawing of a cigarette-smoking hag with yellow teeth and nails.

Yet the hag is skinny, one of the not-so-subliminal ideas behind the naming of Virginia Slims or American Tobacco's Silva Thins. Some women still say they smoke so they won't "reach for a sweet" or grab a fistful of fries.

Then there's the flower-child packaging of Liggett's Eve. In the '90s, Eve may appeal to the 14-year old girly-girl, but the inner girl of the average 18-year-old woman prefers Marlboro. Or maybe a cigar.

"Do you offer a cigar on the first date?" asks a Don Tomas ad. "Go ahead. Make a statement," is the recommendation, edging the cigar maker just past the Chesterfield's 1927 "Blow some my way."

For a gutsier approach turn

to Las Vegas, where a stunning cigar-smoking brunette appears on an ad soliciting visits to Bally's Casino. Today's version of a "sporting girl?" Maybe. Except don't use the word, "girl." Don't assume "sporting" means anything illegal. And remember that some women do smoke cigars.

Bally wants to be where "particular people congregate," to steal the old Pall Mall phrase. In today's advertising lingo, cognac, cigars and wild, wild women define sophistication. The image of health is a cigarette thing. The cigarette is the accouterment to riding through red-rock canyons, schussing down snow-glistened slopes, sailing on waters too blue to be true.

Smoke Eclipse?

In 1988, R.J. Reynolds test-marketed a cigarette with an interior carbon rod that heated tobacco instead of burning it, thus eliminating many hazards of smoke. The consumers who tasted this smokeless cigarette were underwhelmed. But the very idea of it set American anti-smoking groups aflame. They vowed to persuade the Food and Drug Administration that the new tobacco stick was a drug-delivery system masquerading as a cigarette. And so the bell tolled for the infant brand named Premier.

Now the Reynolds people are back again, with a better crafted and better-tasting semi-smoker, according to "Smokescreen" author Philip Hilts. "But it's hard to tell when it's lit and when it's not, which could be irritating to smokers," he said. The invention is called Eclipse.

This experimental cigarette appears safer on so many counts that it could be heaven-sent. In a letter to the FDA asking for oversight of Eclipse, the noted anti-smoking scientist, Dr. John Slade, acknowledged that Eclipse "substantially reduces the levels of many toxins in the aerosol [smoke] a consumer ingests." He added that "Eclipse is different [from other cigarettes for which health claims have been made] only in that it might be able, in a limited fashion, to actually deliver on the hype." In short—-reduced risk of emphysema and cancer is likely.

But Eclipse has a slightly higher carbon monoxide content than old-fashioned varieties, which might mean trouble on the heart-disease front. Slade and his ilk worry about this, as well as other "unintended consequences," including fetal impact and the effects of long use. They also warn that Eclipse will attract youngsters because it leaves no telltale odor.

But my concern—and perhaps yours, too—is that we may never get to try it. R. J. Reynolds is testing Eclipse only in Chattanooga and, under different names, in pockets of Sweden and Germany.

Eclipse. Is the name a sign that Corporate Tobacco is running away from its best idea?

'The vexed question: Is tobacco injurious? Opinions have been far as the poles asunder. Truth as usual seems to lie between, undiscovered by the belligerents.'

F. W. Fairholt, 1859

Warning Label

We know more than Fairholt did, perhaps even as much as that first tobacco enemy, James I, sensed. No form of the weed is hazard free. Tobacco has been tied to more diseases than one can shake a snuff stick at. Some have already been mentioned; now it's time to consider the "C" word, cancer.

Meet gene P53. It resides on human chromosome 17 and its job is to help suppress runaway cell growth. A P53 mutation can be the signature of one or more of fifty-two types of cancer—-most of them not smoking connected. That's the good-news puff.

The bad-news bulletin for cigarette lovers is there's evidence that the benzopyrene (BPDE) in cigarette smoke can attach to P53 mutations, leading to lung cancer. One study has found that when healthy lung cells are exposed to BPDE, the resulting damage looks the same as that seen in many malignant tumors. This could mean that BPDE creates the P53 failure that allows cancerous cells to run wild.

Most smokers didn't need the BPDE news to convince us that smoking can spur lung cancer. We believed the truth of statistical links even if Big Tobacco pretended it didn't. So how big is the risk?

A damaged P53 gene is implicated in half of all malignant lung tumors. But only fourteen percent of smokers develop lung cancer, suggesting that vulnerability may be inborn.

Cancer of the lung isn't smokers' only worry. It's hard to be aerobically fit if one is chronically short of breath; worse still to be part of the canned-oxygen emphysemic set. Tobacco intake is linked to vascular and heart disease, too. Smoking contributes also to other cancers. And—surprise!—nicotine is addictive. The Liggett Group made headline news in 1997 when it became the first cigarette maker to admit this truth in negotiating a separate peace with states suing to recoup costs incurred treating smokers.

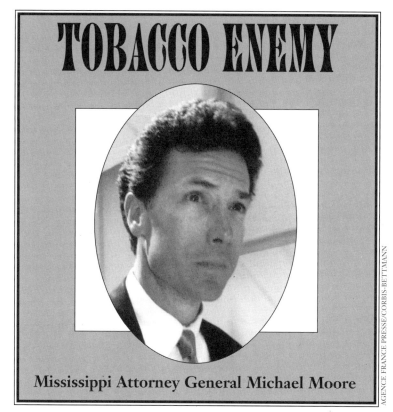

Mississippi Attorney General Michael Moore

BIG TOBACCO PAYS FOR ITS SINS
(And So Will You)

Soon after Liggett's admission, Corporate Tobacco was whacked by a U.S. federal court decision affirming the right of the Federal Drug Administration (FDA) to call nicotine a drug. This enables those guardians of the public health to dictate what goes in cigarettes and how, or even if, tobacco products can be sold.

Who is afraid of the FDA is obvious: Big Tobacco is. And, truth be told, so am I.

In 1997, American tobacco companies also feared the states and cities who were lined up like vultures to sue them for the cost of caring for sick smokers, a cost that may total $50 billion annually. Leading the charge was Mississippi Attorney General Michael Moore.

The tobacco majors—Philip Morris, R.J.R. Nabisco, British American Tobacco (BAT), Lorillard—blinked, offering to negotiate a settlement. Moore and other tobacco foes squeezed out of Big Tobacco a $368.5 billion superfund offer.

That's not tobacco seed. And money was just part of the proffered settlement, which mandated that Big Tobacco knock the Marlboro Man off his horse, bury Joe Camel in the sands of time—consign to the ashes all appealing tobacco images, human and cartoon. Billboard ads, as well as cigarette vending machines, were to go with them.

While most of the moolah would be assigned to settle states' health-cost claims, some $60 billion was tentatively designated for public health funds, anti-tobacco advertising and quit-smoking programs. Penalties were laid on if the industry failed to deter kids from taking up smoking.

What would Big Tobacco get in return? Immunity from future class actions, caps on superfund awards to individuals and

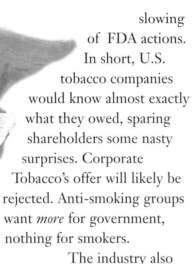

slowing of FDA actions. In short, U.S. tobacco companies would know almost exactly what they owed, sparing shareholders some nasty surprises. Corporate Tobacco's offer will likely be rejected. Anti-smoking groups want *more* for government, nothing for smokers.

The industry also faces costly blows in Latin America and Western Europe, where smoking has declined. In England, class action suits against tobacco companies were recently allowed. But profits are soaring for British and American cigarette makers in Asia, the Mideast and Eastern Europe, where more people demand western brands.

Current anti-tobacco hysteria can be measured by how much the tobacco business is prepared to surrender. Almost any deal between the U.S. government and Big Tobacco is a cheat because government already collects far more in tobacco taxes (an average fifty cents per pack) than it spends (an average thirty-five cents per pack) to care for the tobacco-ill.

Taxes aside, government spends less on smokers than nonsmokers. One reason we're a cheap date is that, statistically-speaking, we go home early. We say good night before certain costly geriatric conditions nab us; we collect social security for a briefer time. Some smokers find solace in this: the five or six years we risk are the last ones.

Smokers have not been party to any of the talks over tobacco deals. But it's likely that any version that survives will punish us.

Big tobacco will recoup at least part of any superfund expenditures by the oldest method known to commerce: raising prices.

It's also a near certainty that the protection rackets run by both government and health insurers will attempt to further cash in. Both tobacco taxes and insurance premiums paid by smokers will zoom up. Outrageous yes, but hypocritical as Uncle Sam may be, the United States has a way to go before matching the greed of tax collectors abroad.

The English and Canadians pay six times as much tax per cigarette pack as Americans do, the Germans four times more, the French and Italians pay triple the U.S. rate, and the Japanese pay double. At least on Japan Airlines' 747s, special seats called *kitsuen senyo* are liberated for those who like to light up with their sake.

Rapacious tobacco taxes have spurred international cigarette smuggling on a huge scale. Such trade could become an American staple, too, if Uncle Sam insists on unreasonably-low cigarette nicotine levels and continues to shoot up taxes.

The $4 pack of cigarettes is coming soon to your neighborhood store, if you can find it. The shop itself may be wrapped in plain brown paper, if tobacco enemies prevail. The health Nazis' goal is forcible withdrawal—ours.

They jubilantly proclaim that if we pay more we'll smoke less. They insist our personal habits are their business. Tobacco foes are

Reynolds Redux

As the grandson of Richard Joshua Reynolds, Patrick Reynolds knows both real fortune and misery. His grandfather, a lifelong tobacco chewer, died of pancreatic cancer. Patrick watched his own father, a heavy smoker, suffer from emphysema and finally fold at age fifty-eight.

Nonetheless, Patrick was a smoker, sticking loyally to the brands of the company his grandfather had founded. He quit, in 1986, when he was thirty-five, after smoking for seventeen years. In "The Last Puff," a collection of quit-stories, Reynolds told his. He was a pack-a-day man who liked the taste of tobacco. His conscious desire to stop warred unsuccessfully for years with his subconscious knowledge that "smoking feels good."

Before he succeeded he was a Colossus of Failure, having flunked Cold Turkey, hypnosis, Smokenders, acupuncture and electric-shock. He went through one program five times. If smoking was a felony instead of a social misdemeanor, his recidivist rate would have had him locked up for life.

Reynolds admits that when a friend lights up a cigarette, he still "really, really" wants one. The tobacco heir has put his money where his desire is, starting Smokefree America.

twisting laws so that the purchase and enjoyment of a legal product is difficult, shameful and seems illegal.

Cigarettes are target number one. The cigar has a partial reprieve only as long as its current chic lasts. Prohibition is a territory already visited. But to realize the futility of attempting to ban what has pleased so many for so long, one needs a passing acquaintance with the wider world. Too many anti-tobacco crusaders know no story but their own.

CAN YOU GET YOUR MONEY BACK?

Don't count on it.

The American tobacco industry has yet to ante up for individual tobacco injury, despite headlines announcing large jury awards to cancer victims who smoked.

One headliner was Floridian Grady Carter, who sued BAT subsidiary Brown & Williamson after he contracted lung cancer. B & W are the producers of Lucky Strike, which Carter took up at age seventeen, and to which she remained true for forty-four years. The jury, able to scrutinize B & W documents leaked by a whistleblower, concluded in 1996 that the company had known since 1963 that nicotine was addictive and that tobacco generally does a body no good. A jury awarded Carter $750,000, a decision B & W is appealing.

In 1997 a different jury, sitting in the same box as the jury who had decided to compensate Grady Carter, heard the case of the family of Jean Connor versus R. J. Reynolds. Connor had died of lung cancer at age forty-nine, after smoking for thirty-four years. Norwood Wilner, the same lawyer who had represented Carter, argued that Jean Connor had been sold "defective" products, namely Winston and Salem, and that Reynolds had been "negligent" in not warning the public earlier about their danger. This jury decided RJR didn't owe the litigants a dime.

Some years ago, an award to the heirs of New Jersey smoker, Rose Cipollene, collapsed after Philip Morris and Liggett appealed it. Although the jury lashed out against the makers of Mrs. Cipollene's favorite brands, it concluded that she was eighty percent responsible for her lung cancer. She started smoking before she knew cigarettes were bad for her, but she didn't quit after package warnings appeared.

Comparable cases crowd court dockets across the country. Big Tobacco is now less sure of winning. Its negotiations with government over a compensatory superfund propose limiting payouts to victors of personal-injury lawsuits to $1 million apiece.

I imagine that one of these days, a cancer victim who quit smoking soon after 1965 when warnings materialized, is going to have a day in court—and that's going to be interesting. Why didn't his or her lungs bounce back rosy?

The nicotine-faithful are dreamers, not schemers. Call it love. Call it addiction. Whatever it is, smoking is not a get-rich-quick gimmick for those who do it.

Gender Zap

Men smokers outnumber women smokers in most of the world, although the gap has nearly closed in the United States and Britain, where a little less than one-third of all adults are nicotine fans. In France, half of all men light up, joined by one woman in three. In Japan and China, nearly two-thirds of the men smoke, but only ten per cent of women have the habit.

'Commit it then to the flames: for it can contain nothing but sophistry and illusion.'

David Hume, 1748

Illusions We Lived With

Scads of whoppers waft through Tobaccoland, including those we smokers tell ourselves. Not every smoking assertion that sounds self-interested is a big black one, or even a little white one. Some are just dumb.

Not until 1997, did cigarette kings, such as Philip Morris CEO Geoffrey Bible and RJR's Steven Goldstone, publicly admit that smoking could lead to fatal disease.

DEPARTMENT OF WISHFUL THINKING

"Tobacco is undoubtedly a poison taken internally. So are coffee and tea. Even if smoking is indulged into excess, the habit never kills."
W.A. Penn, 1901

"We accept an interest in people's health as a basic responsibility, paramount to every other consideration in our business."
Tobacco Industry Research Committee, 1954

"If industry leaders really believed that cigarettes cause cancer, they would stop making them."
Parker McComas, Philip Morris CEO, 1954

"The risks of smoking do not appear to differ significantly from lots of other calculated risks to which modern man exposes himself."
Charles Cameron,
American Cancer Society, 1954

Vivien Leigh and Clark Gable in GWTW. Springer/Corbis-Bettmann

Scarlett O'Hara Award
To would-be quitters who decide: "I'll think of it all tomorrow."

"People who smoke don't give a shit about their health."
Anonymous R.J. Reynolds executive, 1976

"I think there's a great deal of doubt as to whether or not cigarettes are harmful."
Helmut Wakeham, Philip Morris, 1976

"There isn't a mounting weight of evidence [linking smoking and illness], there's a mounting weight of propaganda."
Robert Walter, American Tobacco CEO, 1963

"We believe there is no [health] connection or we wouldn't be in business."
James Bowling, Philip Morris executive, 1963

"We don't know of anything that makes a cigarette unsafe."
David Fishel, R.J. Reynolds, 1985

"It is certain that a substantial portion of the lung cancers that occur in nonsmokers are due to ETS [secondhand smoke]."
Everett Koop, former U.S. Surgeon General, 1986

"Meat kills more than smoking."
Real-eat Encyclopedia of Vegetarian Living

FLATTERY WILL GET YOU EVERYWHERE

"Most cool people smoke. Because the kind of people who smoke tend to be stressed, that is to say they are the people who tend to be active and doing interesting things."
B.J. Cunningham,
Enlightened Tobacco, 1997

A THOUSAND TIMES NO

"I believe nicotine is not addictive."
Edward Horrigan, RJR Nabisco, 1994

"I believe nicotine is not addictive."
Thomas Sandefur, B & W CEO, 1994

"And I, too, believe that nicotine is not addictive."
Ross Johnson, RJR Nabisco CEO, 1994

The "statistical data has not convinced me that smoking causes death."
Andrew Tisch, Lorillard CEO, 1994

"Philip Morris does not measure nor independently control for the level of nicotine in our products."
William Campbell, Philip Morris CEO, 1994

"I'm not certain whether it's addictive."
Presidential candidate Bob Dole, 1996

IF ELEPHANTS COULD FLY

"Just one puff won't send me back."
Every recidivist in smoking history.

'I left the cigar lying there, on the couch.'

Groucho Marx, 1978

Smokers' Luck

Smokers are gamblers at heart. Some who play with matches get burnt in funny ways. Lady Luck beams on others because they smoke. A few are stung by the slings of outrageous fortune merely because they hang out with puffers. Luck happens.

GO DIRECTLY TO JAIL

In 1908, New York City's Board of Aldermen passed a law against women smoking in public. A cop, whose name has been forgotten—-but we know the type—-lost no time in arresting 29-year-old Katie Mulcahey when she lit up a cigarette. Mulcahey did not go meekly in the arms of the law. "No man shall dictate to me," she declared.

Two weeks later, the law was vetoed by Mayor McClellan, an unsung hero of the rights of both smokers and women, who did not yet have the vote.

TOO HOT FOR GROUCHO
(As told to Charlotte Chandler by Groucho Marx)

Groucho: "So I picked up this girl one day, and she was pushing a baby carriage. I spoke to her and I says, 'You're a pretty girl. Are you married?'

"She says, 'No, this is my sister's baby.' She was lying; it was her baby.

"She took me up to her apartment, and I was smoking a cigar. Suddenly, I hear the sound of footsteps. So I run to the closet but I left the cigar lying there, on the couch. And this guy comes in. He says, 'There is a man in here.' She says, 'There is not.'

" 'What is the cigar doing here?' He looks into the closet.

" 'If I find the son of a bitch,' he says, 'I'll kill him.' And he felt in the closet, but there were a lot of clothes in there, and he didn't feel me. The minute he went into the kitchen I jumped out the window."

Chandler: "What floor were you on?"
Groucho: "The first."

Hello, I Must Be Going: Groucho and His Friends (1978)
by Charlotte Chandler

A MATTER OF HEALTH

Sadi Ranson, a young Englishwoman living in Boston, decided to quit smoking in August, 1996. She smoked Parliament or what she and her brother call, "Nervous 100s."

"It wasn't that hard. I'd done it before," she said.

The next day she felt awful and then she felt even worse. Three weeks later, her head still ached, she was nauseous and she had a temperature. She took herself to a medical clinic, seeking relief from her withdrawal symptoms. Clinic staff called an ambulance to ship her to the emergency room of Brigham and Women's Hospital.

E.R. medics couldn't figure out the cause of her misery until a doctor ordered a spinal tap.

The diagnosis was meningitis. Ms. Ranson was quarantined in Intensive Care, where her life fluttered in the balance. With proper treatment, she turned the corner. "Your immune system was shot to hell," a resident told her when he visited her bedside. "Why should that be?" Have there been any drastic changes in your lifestyle recently?"

"I gave up smoking," Ms. Ranson

replied.

"The doctor said that the change had so shocked my body that I had surrendered all my defenses," the Englishwoman recalled a year later. "My immunities had been knocked out."

After her release from the hospital she considered the throttled state in which quitting had left her. "I thought the hell with it. Smoking is one of the few things I really enjoy."

Sadi Ranson and her Nervous 100s got back together.

HE LIT UP THE HOUSE

On Saturday night, August 13, 1994, Frank Oliveto, an orthopedic surgeon from Belle Terre, N.Y., fired-up a cigar at a twenty-one table in Atlantic City. But the rules had changed, and the dealer regretfully told Dr. Oliveto that he was at a no-smoking game. Leaving his wife, Margaret, hostage to fortune at the blackjack table, Oliveto, cigar clamped between his lips, dragged over to the slots.

He procured $100 worth of dollar tokens and proceeded to stuff a progressive slot machine, $3 a go. The doc was about forty-five bucks down when a blitzkrieg of lights and chimes froze the gaming room. Frank Oliveto had won the largest jackpot ever: $8.5 million.

SNUFF OUT

On February 5, 1712, the Dauphine of France, suffering from what she believed was a toothache, accepted a box of Spanish snuff from the Duke of Noailles. She placed the box on a table in her private boudoir. The princess also enjoyed pipe smoking and tobacco chewing, pleasures she kept secret from her father-in-law, Louis XIV.

The Spanish snuff did not soothe her pain. And neither her pipe nor chew (nor two bleedings) eased the towering migraine and fever that sent her to bed. Before the week was out, the princess died.

The snuffbox was missing from the

dauphine's boudoir at the time of her death and was never again found. Some whispered of poisoning: the contents of the lost snuffbox were extremely suspect.

The mystery of the dauphine's death was never solved. But Spain developed a rep for snuffing out its enemies with adulterated—you got it—snuff.

MORMON WHISTLE STOP

Business executives John Lynch and Ernest Bamberger had just finished their after-lunch cigars at the Vienna Cafe in Salt Lake City on an afternoon in 1923, when deputy sheriffs Michael Mauss and John Harris strode in. Spying the overflowing ashtray on the businessmen's table, they seized the evidence and arrested Lynch and Bamberger for the crime of smoking in public. They also nabbed Edgar Newhouse whom they'd caught *flagrante delicto* with a cigarette, according to an *American Heritage* summary of the black day.

The three criminals were marched down Main Street to the county jail, where they were booked and held, pending the posting of bail.

The resulting uproar, orchestrated by the Chamber of Commerce, included the blasting of steam whistles throughout the city to protest the law banning cigarette sales to adults and smoking in restaurants. The Mormon-dominated Utah legislature rescinded the law, but a similar one is in effect today. Cigarettes may be sold but not smoked in public.

A BUTT ABOUT IT

Firemen, answering a call on Figure Eight Island, N.C. in April, 1997, found a lovely vacation home engulfed in flames. When the smoke cleared, authorities said they suspected the fire had been ignited by a cigarette butt, accidentally left lit. The house belonged to Andrew Schindler, president of R. J. Reynolds Tobacco.

A DAMPENER

Sir Walter Raleigh was enjoying a pipe in his dressing room when a palace servant entered with a basin of water for his toilette. The startled servant, who had never before seen smoke streaming out of anyone's mouth, assumed he was on fire and emptied the basin on Sir Walter's head.

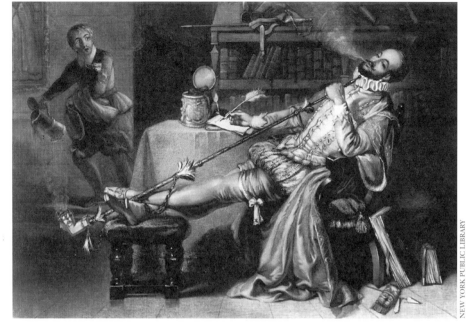

BURNED IN REAL ESTATE

André-Francois Raffray, age forty-seven, thought he had the bargain of a century in 1965 when he bought Jeanne Calment's large apartment in Arles, France. The deal was he would pay Mme. Calment $500 a month until she died, and then take possession of the flat. Since Mme. Calment was ninety years-old at the time, Raffray didn't mind the wait.

Jeanne Calment had a secret weapon, though: she smoked. Thirty years later, when Raffray died, Mme. Calment was a healthy 120-year-old, who recently had released a C.D., offering reminiscence set to rap. At the time of his death, Raffray had paid out $184,000 for the apartment, and his heirs inherited the monthly obligation. Legal title to the flat finally came to the Raffray family when Mme. Calmet died in 1997—at the grand old age of 122.

C'EST LA GUERRE, ONE

In this Napoleonic war story, Jean Nepoma Bouffardi, a corporal in the grand army of France, was pulling on his pipe at Friedland. The next day his arm was found on the battlefield, his fingers still clutching his pipe.

C'EST LA GUERRE, TWO

Marshall Oudinot, wed to his pipe, led Bonaparte's troops to victory in several engagements. Although Napoleon famously choked on tobacco smoke, he rewarded Oudinot with a meerschaum in the shape of a mortar on a carriage, encrusted with thirty-thousand francs' worth of diamonds.

SMOKE FOR LOVE

Club singer Sandra Purdy of Cocoa Beach, Fla., was down on her luck in 1994, suffering from Lupus disease. A lifelong smoker, she accepted a job at the Plum Lounge, securing petition signatures at fifty cents a pop for the National Smoking Alliance. Ms. Purdy moved in on a guy smoking at the bar, sitting next to another man who was only nursing a drink. "I chatted up the smoker to get his signature and ended up falling in love with the other guy," she said, summing up the story. She and the "other guy" remain an item. Sandra Purdy still smokes; he still doesn't.

EVERY INCH A PRINCE

The prince of Wales visited Canada with his royal entourage in 1859. That far from the disapproving eye of his mother, Queen Victoria, he felt free to light cigar after cigar. The party was touring a desolate, wind-ripped prairie when the future King of England and some of his pals decided on time-out for another smoke. Finding they had only one match left among them, they drew lots for the solemn responsibility of lighting it and spreading the flame. The prince won.

Tented by his companions, the prince struck the match and succeeded. Decades later, Edward VII characterized the moment as the most nervous and exciting of his entire life.

WINNERS ALL

The winner of an 1890 farting contest in London attributed his victory to having taken a smoke enema before appearing on stage.

Dazzling as his performance was, the Londoner was easily outclassed by French music hall star Pétomane (né Joseph Pujol), who thrilled adoring *Moulin Rouge* audiences early this century by turning his back on them. One signature act was smoking a cigarette in a rubber-tube holder inserted in his anus. Perfect sphincter control allowed him to inhale and exhale with grace. Le Pétomane was an *artiste*, not a streaker. He performed in formal attire, although there was a discreet hole in the seat of

Le Pétomane, in performance, also smoked through an unusual orifice.

his pants.

Replacing the cigarette with a small flute, he played *Au Clair de la Lune*, not usually thought of as a smoking song. Le Pétomane commanded twenty thousand francs per engagment, outdoing the other Paris sensation, Sarah Bernhardt, who settled for eight thousand.

Cut to Transylvania, 1996, where record-book aspirant, Stefan Sigmond, triumphantly, albeit conventionally, smoked eight hundred cigarettes in six minutes. "I have an awful taste in my mouth, but I'm sure it will go away, " Sigmond told the press.

NEAR DEATH IN VENICE

Lee Travers was on a *vaporetto* heading for her hotel. With her was her ten-year-old son, Puck. Done with her cigarette, she tossed it in to the Grand Canal—or so she thought. "Mom," whispered an aghast Puck, "your cigarette landed in that lady's hair."

The mother followed her son's glance. Sure enough, there was her sizzling butt sitting atop the elaborate, upswept blue-gray hairdo of a Venetian *signora*. "I had no idea what to do," Ms. Travers recalled. "Should I go up to her and say, 'Excuse me, madam, but my cigarette is in your hair.' And how would I say that in Italian? I was mortified, frozen, and any second her hair would be in flames."

The American tourist and the unknowing *signora* were saved by a gust of wind that blew the butt out of the latter's hair into the canal.

DRESSED TO KILL

Georgia Derrick Dudley maintained a most proper Virginia household, her daughter remembered a half-century later. "So one time when company came unexpectedly, my mother put her lit cigarette in her apron pocket as she went to answer the door. I followed her and saw the smoke curling up."

The visitor lingered so long the apron began to smolder. Both guest and hostess were too polite to mention this until the daughter took action. She patted out the fire which ruined both apron and underlying dress. "It was very embarrassing," the daughter said, "I can't think of any time more embarrassing I ever saw."

CIGAR TO THE RESCUE, ONE

Italian patriot Guiseppi Mazzini, in London exile, was puffing a cigar in his rooms when assassins burst on the scene.

"Take a cigar, gentlemen," Mazzini greeted them. "I believe your intention is to kill me," he added, offering round his box.

The hired killers hesitated. How could they deny their victim enjoyment of his last smoke, or insult his hospitality by refusing to join him? They smoked with him—-and left, mission unaccomplished.

CIGAR TO THE RESCUE, TWO

Journalist Thomas Goltz was in an awkward spot. He was pinned to a runway in Chechnya by an excitable security agent, named Ilkhani, who held a cocked pistol in Goltz's ear, as the plane Goltz needed to catch hummed for take-off.

"What the hell is going on here?" screamed the Russian pilot out the cockpit window.

As Goltz wrote the story in a 1997 *New York Times Magazine* essay, his escape route was clear. One Cuban Partagas or "Uncle Fidel," for Ilkhani, one for the airplane steward who re-opened the door, and two for the pilot who whisked him to Moscow. Mellowed, the pilot instructed Goltz, "Bring me a box next time, my brother, and you can hijack me to the moon."

CIGARETTE SOLACE

When the *Formidable* sunk in January, 1915, "Captain Loxley stood on the bridge calmly smoking a cigarette," according to a *New York Globe* account based on survivors' reports.

The newspaper observed, "We also hear of other heroes who go to their doom, lipping a cigarette between their teeth. It is never a cigar or a pipe, always a cigarette."

GET A MOVE ON

Alex Sini, stuck in peak-hour Sydney, Australia traffic, decided to counter frustration by rolling a cigarette. "I thought I had the technique down pat, " Alex Sini recalled. "I put my head down, licked the paper, then smiled at myself for being so talented. I stuck the masterpiece in my mouth and lit it up.

"A wad of tobacco burst into flames just as the traffic started to move. Car horns started going off and there I am with a head full of frizzling red hair."

The horns just kept blaring. Rush-hour drivers in Sydney accept no excuses. Traffic stops happily for no man, woman or smoking disaster. So Sini frantically patted out the flames as a thousand car trumpets serenaded the misfortune.

TRULY TWISTED

Margaret Finch, "Queen of the Gypsies," expired in 1740 at the age of 108, after a lifetime of enjoying a pipe. She died peacefully at home in Norwood, England, sitting in her usual pipe-smoking position.

Her usual position was the problem. According to the obit: "From a habit of sitting on the ground, with her chin resting on her knees, her sinews became so contracted, that she could not rise from that position.

"After her death, they were obliged to enclose her body in a deep, square box." However unorthodox the coffin, "a great concourse of people attended" the queen's funeral.

CRIME WAVE

Nineteen ninety-six was a bad year for Smokers versus The Law, although there are no official statistics. Among the "curiouser" stories coming from the American Wonderland:

In Carson, California, cops busted a dollar-a-game pinochle party that retirees had peacefully played daily for twenty years. The complaint that led police to the clubhouse doors wasn't gambling, or an emergency coronary attack, it was smoking.

In Stevenson, Md., college president Carolyn Manuszak was fined $1,312 for smoking in her private-office bathroom. After being summoned to campus by a cowardly college employee, who insisted on anonymity, investigators followed smoke signals from the open bathroom window to the perp.

BUT DID HE STOP SMOKING?

Silas Leite de Silva and his wife began 1997 with a blazing argument. The New Year's Day quarrel took place in their Campinas, Brazil home. What set Senora de Silva fuming, according to police, was her husband's insistence on smoking inside.

The fight, which exhausted Senor de Silva or at least put him off guard, fired-up his wife to violence. She grabbed a knife and unmanned—Bobbittized—her spouse.

BUT DID SHE STOP SMOKING?

After fifty years of cigarettes, Sally Thomas of Chicago is still pedaling—a bike. She took to her wheels, the *Chicago Tribune* reported, after her husband of forty-three years, Col. Richard J. Thomas, went to federal court on Aug. 21, 1997 to demand that her smoke be declared toxic under the Clean Air Act.

At 6 AM the next morning, Mrs. Thomas slipped out of the stockade she calls home, and successfully fled on her bicycle to avoid an oncoming TV crew. The colonel, an ex-smoker, later claimed he acted out of love. "I'm just trying to take care of her," he said.

Keep on pumping, Sally Thomas. Don't even think of turning back.

'Time, time is all I lacked to find them, as the great collectors before me.'

Keith Douglas
On a Return from Egypt, 1944

Tobacco Treasures To Die For

Tobacco burnishes a myriad of objects, from pre-Columbian artifacts to Bakelite accessories and Art Deco advertising posters. Before there were vending machines, there was the penny-tobacco box of 19th-century taverns. The smoker slipped in a penny and the box sprang open. Another rarity is the 17th-century ember tong, a silver implement fashioned especially for lifting charcoal chips from a juniper fire to a pipe bowl.

Several historic houses and a particular Renoir are among the glories of tobacco's legacy. So is the Arents Collection in the New York Public Library, an archive of tobacco-related manuscripts and art, covering four centuries of leaf history. Many museums, including *Artes decoratifs* in Paris and London's Victoria and Albert, have dazzling snuffbox collections. You won't find a smoking artifact section at most antique shops or flea markets, yet tobacco treasures are nearly everywhere you look.

HOUSES TOBACCO BUILT

Monticello (1), designed by Thomas Jefferson, is arguably the finest historic home in America. Jefferson insisted on paying off his debts to British tobacco merchants after the Revolutionary War, although King George III's soldiers had destroyed his crop. His honor combined with his French shopping habits and his continuous redesign of Monticello, which took forty years to build, almost bankrupted him.

The Byrd Manor House (2) is one of the great James River tobacco-plantation houses.

The marble Schinasi Mansion (3), completed in 1909, sits on a Manhattan height overlooking the Hudson. Morris Schinasi emigrated from Turkey and made a fortune importing Turkish leaf to satisfy the American craving for exotic cigarettes. He hired the architect who had designed Carnegie Hall to build his New York family residence. Unlike the tobacco trophy-homes which have become museums, Schinasi is still privately owned.

The drawing room of Reynolda House (4) in Winston Salem, N.C. Reynolda is the farmhouse that R. J. Reynolds built for himself in 1917. The house was occupied by Reynolds' descendants until 1963 and is now an art museum.

HIGHER EDUCATION

The chapel at Duke University in Durham, N.C. Washington Duke endowed small Trinity College in 1896, stipulating that women be admitted "on equal footing with men." In 1924, his son, James Buchanan Duke, provided the tobacco fortune that transformed the school into nationally-renowned Duke University.

ART APPRECIATION

"Monsieur Fournaise," 1875, by Pierre-Auguste Renoir.

"The Card Players," 1892, by Paul Cézanne.

"Al Capone," 1972, by LeRoy Neiman.

Snuffboxes are big among individual collectors as well as museums.

THE JOKE'S ON US

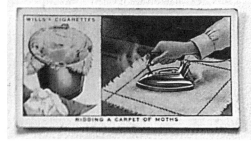

Collectible cigarette cards for every taste: humor series, tips on flowers, housekeeping hints.

The art is in the artlessness of this old lithographic advertisement.

One of a pair of bronze spittoons, cast from Civil War materiel, which once graced the Grand Army of the Republic room at the Chicago Public Library.

PRECIOUS PIPES

In the cigarette-burning, big-band era, a stylish accommodation was made for old-fashioned smokers: the art deco pipe.

This pipe of hand-wrought sterling silver with a meerschaum bowl was valued at a hot $25,000 in 1941 when it appeared in an ad for Lucky Strike plug tobacco.

Silver and porcelain Chinese pipe, just big enough for a dash of tobacco or opium.

This antique hand-painted porcelain and wood pipe, from Austria, is a tobacco-lover's treasure

STAR VALUES

John F. Kennedy's humidor was sold by Sotheby's in 1996 for $574,500 to Marvin Shanken, publisher of "Cigar Aficionado." A losing bidder was comedian Milton Berle, who had purchased the Dunhill walnut humidor thirty-five years earlier for $800. The plaque on the world's most expensive humidor reads, "To JFK. Good Health. Good Smoking. Milton Berle."

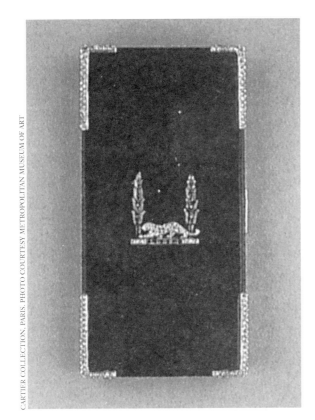

This exquisite, bejeweled black-lacquer-on-gold cigarette case was made by Lavabre for Cartier in 1928.

TOBACCO TAPESTRY

Turn-of-the-century freebies have acquired modest value. A quartet of queenly tobacco silks (shown this page); a unique jacket (shown on next page) crafted from silks featuring ballplayers, actresses and butterflies.

CACHET ASHTRAYS

THEATER PIECES

The sterling characters of the cigarette case and cigarette box resonate drama. The case was presented to Cole Porter as a memento of "The Man Who Came to Dinner;" the autographed box was given to director Tully Marshall by the London cast of "The Adventure of Lady Ursula" on Oct. 11, 1898.

TOBACCO TRIFLES

Something old,
something new:
images of Bogie, Camel
and the Lucky logo have been recycled
as buttons, while Black Cat comes back as a magnet.
Such finds cost little more than a pack of cigarettes.

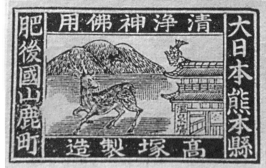

GLOBAL FIRE WORKS

Yesteryear's throwaways have antique allure today, a theory illuminated by this around-the-world array of old matchbox labels.

'The nature of men is always the same; it is their habits that separate them.'

Confucius, circa 500 B.C.

Charmed Leaf

The "Indian weed" is a boon to humankind in more ways than most of us realize.

MATCHLESS EXPERIENCES: NOVEL TOBACCO USES

WEAR IT
A panatela peeking from the lapel pocket of a white suit epitomizes summer chic. Stick it behind your ear and it screams power-wonk.

DRINK IT
Follow the elegant example of Guyana's First Tribe, and take tobacco tea. Or, when in Seville, mix an ounce of tobacco with a gallon of Spanish wine for a drink heartier than Sangria.

Warning: Tobacco as a thirst-quencher seems full of risk.

BRUSH YOUR TEETH
According to folkways, cigar ash mixed with poppy oil makes a fine toothpaste.

DECORATE YOUR GUMS

This tip comes from the Kogi of Colombia who disapprove of smoking and only use organic cosmetics.

> ### TOBACCO GUM PAINT*
> Boil tobacco for several days.
> Add manioc starch (tapioca) to thicken tobacco syrup.
> Store result in a hollow gourd.
> Use sturdy stick to apply plaster to gums.
>
> *Anthropologist Uscategui Mendoza who once attended a Kogi fashion show cautioned that gum painting is addictive.

EXORCISE WITCHES

Sprinkle tobacco seed by your bed to assure unhaunted sleep. Virginian Belle Boswell says her grandmother told her, "The witch has to count every seed so she can't mess with you."

CLOCK IN

Certain American tribes reckoned time in pipe-smoke lengths. "I was one pipe at it," said a Native American, requesting his wage.

MEASURE DISTANCE

Some South American country people describe the length of a journey in tobacco; i.e., the walk to the next village is a two-leaf-chew.

DRIVE CATTLE

After a hard night on the trail, Texas cowboys rubbed tobacco juice in their eyes to keep awake.

SAY CHEESE

The Southern Growth Policies Board reported in 1996 on several uses for tobacco derivatives, including nutrition supplements, antiseptic mouthwash and cheese-making enzymes.

HEALING HIGHLIGHTS

Tobacco can cure anything and everything that ails you, if you believe centuries of medical advice.

In 1527, Nicholas Monardes prescribed "nicotine oyntment" for cleaning and knitting together wounds.

> **NICOTINE OYNTMENT**
> Take one pound fresh tobacco leaves.
> Pound them.
> Mingle them with new resin, common oil—-three ounces of each.
> Boil together.
> Add three ounces of Venetian turpentine.
> Strain though linen cloth.
> Keep it in pottes for your use.

Sixty years later, Gilles Everand of Antwerp, recommended sticking tobacco in

Doctor's Orders: An unhappy Native Brazilian patient is about to be purified with cigar smoke in this 1555 Andre Thevet woodcut.

your ear as a cure for deafness. "Apply with a drop of worm oil," he told patients. He also recommended tobacco for tetanus, tonsillitis ("pains of the throat overcooked by rheume"), epilepsy ("falling sickness"), and syphilis ("French pox"). For fevers, Everand ordered a tonic of oil of brimstone mixed with aged tobacco.

Taking their cues from Native Americans, western physicians treated ulcers, kidney stones and other internal ailments with tobacco smoke. The clyster smoke enema was discussed in the chapter on style. A "winning" benefit of the smoke purge appeared in *"Smokers' Luck."* The original technique, pictured on page 263, was not gentle.

Other tobacco cures tasted awful. In 1685, an Englishman, serendipitously named Ashole, recorded in his diary: "I was much troubled by my teeth in my upper jaw, on my left side for a week. Then I held pills in my mouth made of burned alum, pepper and tobacco and I was much eased."

Leslie Hanscom, raised in Kennebunkport, Maine, remembers the tobacco toothache-cure as a country remedy.

UCLA professor Johannes Wilbert has witnessed South American shamans blowing smoke on scorpion stings, and using tobacco for post-tattoo treatment.

Pretty primitive, right?

According to a recent *Newsweek* story on future medicine, tobacco is being looked at as a treatment for a genetic disorder of the spleen, liver and bones. Experiments in its use in new antibiotics and antifungals are also underway.

OF EARTHLY GOOD

A Macuxi shaman in Maloca Bananl, Brazil, washed his mouth with an elixir of tobacco before summoning mountain spirits. The priest was calling on the spirits to save tribal land from intruders hell-bent on gold. This description might apply to the age of the conquistadors, but it paraphrases a recent *New York Times* report.

They came looking for gold and found tobacco. The "Indian weed" has hugely enriched relatively few but supported many. The charmed leaf also adds grace to the earth itself, a role that largely has been forgotten.

Once upon a time, ornamental gardens in temperate zones around the world gloried in the tobacco plant. Earlier this century, ordinary flower growers understood another

tobacco plus. They evicted insects from their flower beds with tobacco juice. They grew roses with tobacco mist.

TOBACCO MIST

Mix contents of one unfiltered cigarette or one fresh cigar with one pint boiling water.
Dip in ordinary teabag (no herbals!) and remove.
Let liquid cool.
Add 1/2 tsp. ammonia.
Add 1/2 tsp. dishwashing liquid.
Mix well.
Water soil with 1/2 cup of the nutrient.
Use the rest of the potion to spray leaves.

TOBACCO JUICE

Empty ashtrays into cauldron of water.
Boil up.
Snuff out fire.
Let water cool.
Strain out butts.
Sprinkle liquid on soil.

BEFORE YOU PUT THAT CIGARETTE OUT

- Blow up a train—as Lt. Dunbar did in *Stalag 17*.
- Fry ticks.
- Burn leeches from your legs on jungle walks.

WHO SAID SMOKING IS DIRTY?

From ashes to godliness—the remains of what you smoke constitute one of the world's great cleansers.

POOR MAN'S PIMM'S CUP

A damp cloth dipped in cigarette ashes will shine up your best pewter mug.

SECRETS OF THE ADMIRALTY

Spruce up the teak trim on your craft with a dab of mayonnaise mixed with the residue of your pipe. Hold the mayo and you still get a decent gleam.

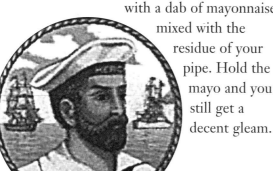

ONLY YOUR HAIRDRESSER KNOWS FOR SURE

A paste made of cigarette ashes and baby shampoo rubs hair dye right off your skin, clothes and the floor.

GENTLEMEN'S CLUB TRICK

You didn't use a coaster under your tumbler and it left a telltale ring on the leather. Here's what the butler knew: dribble water and ashes over the ring, pat with dry cloth, then spread a dab of hand lotion over the offense.

COMING YOUR WAY, MAYBE

The U.S. patent office has recognized the invention of an environmentally-benign rust inhibitor made of tobacco.

YOU CAN SEE CLEARLY NOW
Window fogged? TV screen smudged? Windshield splattered? Liberate tobacco from its wrapper and use it to wipe away the woes of modern life.

BENEFITS OF SECONDHAND SMOKE

- Repels mosquitoes.
- Masks other odors.
- Clears crowded rooms.

Bibliography

Adamson, J.H, & Fulland H.F., The Shepherd of the Ocean: An Account of Sir Walter Raleigh and His Times, Gambit, 1969.
Andrews, Lincoln Clark, Fundamentals of Military Service, U.S. Army, 1944.
Apperson, George, The Social History of Smoking, Ballantine, 1914.
Arber, Edward, ed., English Reprints, 1869 (James I's "Counterblaste to Tobacco").
Bain, John, Tobacco in Song and Story, H.M.Caldwell, 1896.
Barrie, My Lady Nicotine, Hodder & Stoughton, 1890.
Beaumier, John Paul & Camp, Lewis, The Pipe Smoker, Harper & Row, 1980.
Brooks, Jerome, Tobacco: Its History, Rosenbach, 1939.
Burrough, Bryan & Helyar, John, Barbarians at the Gate: The Fall of RJR Nabisco, Harper & Row, 1990.
Carns, Tracy, ed., The Cigar in Art, Overlook Press, 1996.
Cook, Eben, The Sotweed Factor or A Voyage to Maryland, D. Bragg, 1708.
Corti, Count Egan A History of Smoking, George Harrap, 1931.
Crosby, Jr., Alfred W., The Columbian Exchange, Greenwood, 1972.
Dickens, Charles, American Notes, Chapman & Hall, 1842.
Dor-Ner, Zvi, Columbus and the Age of Discovery, William Morrow, 1991.
Dunhill, Alfred P., The Pipe Book, Macmillan, 1924.
Fairholt, F.W., Tobacco: Its History and Associations, Chapman & Hall, 1859.
Farquhar, John W. and Spiller, Gene A., The Last Puff, W.W. Norton, 1990.
Ford, Henry, The Case Against the White Slaver, Henry Ford, 1914.
Hacker, Richard Carleton, The Ultimate Cigar Book, Autumngold Publishing, 1993.
Heimann, Karl, Tobacco and Americans, McGraw Hill, 1960.
Herment, Georges, The Pipe, Simon and Schuster, 1955.
Hilts, Philip J., Smokescreen, Addison Wesley, 1996.
Irwin Margaret, That Great Lucifer, Harcourt Brace, 1960.
Jefford, Andrew, Smokes, Evening Standard Books, 1996.
Josephy, Jr. Alvin M., Five Hundred Nations: An Illustrated History of North American Indians, Alfred A. Knopf, 1994.
Klein, Richard, Cigarettes are Sublime, Duke University Press, 1993.
Kluger, Richard, Ashes to Ashes, Alfred A. Knopf, 1996.
Krogh, David, The Artificial Passion, W.H. Freeman, 1991.
Lissauer, Richard, Encyclopedia of Popular Music, Paragon, 1993.
Nath, Uma Ram, Smoking: Third World Alert, Oxford University Press, 1986.
Neihardt, John G., Black Elk Speaks, University of Nebraska Press, 1995.
Nickles, Sarah, ed., Drinking, Smoking & Screwing, Chronicle Books, 1996.
Penn, William, The Soverane Herb, E.P. Dutton, 1901.
Pollock, Edward, Sketchbook of Danville, E.R. Waddill, 1885.
Redway, George, Tobacco Talk, Brentano, 1884.
Sobel, Robert, They Satisfy: The Cigarette in American Life, Anchor Press, 1978.
Strasser, Susan, Satisfaction Guaranteed: The Making of the American Mass Market, Pantheon, 1989.
Wallace, Willard, Sir Walter Raleigh, Princeton University Press, 1959.
Wilbert, Johannes, Tobacco and Shamanism in South America, Yale University, 1987.
Williamson, Ross, Sir Walter Raleigh, Greenwood Press, 1978.